INFORMATION LITERACY FOR SCIENCE AND ENGINEERING STUDENTS

AF729496

INFORMATION LITERACY FOR SCIENCE AND ENGINEERING STUDENTS

CONCEPTS AND SKILLS

Mary DeJong

BLOOMSBURY LIBRARIES UNLIMITED
NEW YORK • LONDON • OXFORD • NEW DELHI • SYDNEY

BLOOMSBURY LIBRARIES UNLIMITED
Bloomsbury Publishing Inc
1385 Broadway, New York, NY 10018, USA
50 Bedford Square, London, WC1B 3DP, UK
29 Earlsfort Terrace, Dublin 2, Ireland

BLOOMSBURY, BLOOMSBURY LIBRARIES UNLIMITED and the Diana logo are trademarks of Bloomsbury Publishing Plc

First published in the United States of America 2024

Copyright © Mary DeJong, 2024

For legal purposes the Acknowledgments on pp. 323–327 constitute an extension of this copyright page.

Cover image: Sabena Jane Blackbird/Alamy Stock Photo

All rights reserved. No part of this publication may be reproduced or transmitted in any form or by any means, electronic or mechanical, including photocopying, recording, or any information storage or retrieval system, without prior permission in writing from the publishers.

Bloomsbury Publishing Inc does not have any control over, or responsibility for, any third-party websites referred to or in this book. All internet addresses given in this book were correct at the time of going to press. The author and publisher regret any inconvenience caused if addresses have changed or sites have ceased to exist, but can accept no responsibility for any such changes.

Library of Congress Cataloging-in-Publication Data
Names: DeJong, Mary, author.
Title: Information literacy for science and engineering students: concepts and skills/Mary DeJong.
Description: New York: Bloomsbury Libraries Unlimited, 2024. | Series: Bloomsbury Libraries Unlimited series for teaching Information literacy today | Includes bibliographical references and index.
Identifiers: LCCN 2023046625 (print) | LCCN 2023046626 (ebook) | ISBN 9781440878763 (paperback) | ISBN 9781440878770 (ebook) | ISBN 9798216171850 (epub)
Subjects: LCSH: Science–Information services. | Technology–Information services. | Information literacy. | BISAC: LANGUAGE ARTS & DISCIPLINES / Library & Information Science/General | SCIENCE/Research & Methodology
Classification: LCC Q179.96.D45 2024 (print) | LCC Q179.96 (ebook) | DDC 501/.154–dc23/eng/20231220
LC record available at https://lccn.loc.gov/2023046625
LC ebook record available at https://lccn.loc.gov/2023046626

ISBN: HB: 979-8-2161-8858-2
PB: 978-1-4408-7876-3
ePDF: 978-1-4408-7877-0
eBook: 979-8-2161-7185-0

Series: Teaching Information Literacy Today

Typeset by Integra Software Services Pvt. Ltd.

To find out more about our authors and books visit www.bloomsbury.com and sign up for our newsletters.

CONTENTS

SERIES FOREWORD

Imagine receiving a mandatory invitation to a gathering where you do not know anyone. Worse still, it is a formal event with a strict dress code, unfamiliar dining utensils, *and* you will be expected to give a speech. A single misstep could have lasting repercussions on your future social and professional relationships. Now, imagine that you do not have to navigate this terrifying scenario alone. Instead, you are given a handbook full of clear advice on which fork to use, what jokes will win over the room, and precise directions for executing every secret handshake.

In *Information Literacy for Science and Engineering Students*, Mary DeJong offers exactly that; with clarity and warmth, Mary invites her readers into an unfamiliar information world and demystifies the conventions, structures, and resources that define research in science and engineering. Perhaps you are a chemistry major in your first semester. Or, an aspiring civil engineer writing your first technical report. Maybe it has been many, many years since you've entered the field, and you no longer remember the difference between a review article and research article? No matter how you come to this book, we know that Mary's straightforward explanations, concrete examples, and cheeky humor will offer the ideal entry into scientific communication.

We are thrilled to have Mary's book as the inaugural title in the Teaching Information Literacy Today series. This series publishes practical, theoretically informed books for librarians in higher education who are tasked with teaching information literacy.

Over the last decade, librarians have collectively witnessed the transformation of our information landscape. Notably, the ascension of search engines as primary tools for information acquisition, the proliferation of social media platforms as venues for online interaction and entertainment, and the emergent ubiquity of mainstream artificial intelligence. Individuals are casually and almost universally equipped with pocket computers, which diminish our attention spans and inhibit our capacity for focused engagement and deep reading.

This evolving information ecosystem has been further complicated by an escalating influx of disinformation, misinformation, and conspiracy theories. Individuals have been commodified, personal privacy has been infringed upon, and human relationships have been reshaped. Each successive techno-social paradigm shift has irretrievably altered the infrastructures we work within to access, store, and process information—tools integral to both educators and learners. The landscape of information grows ever more complex and volatile, while superficially easier to access.

In teaching information literacy, we help individuals better understand the information systems they navigate and the information formats and genres that move through those systems. The titles in this series represent the topics, approaches, and priorities that help us, as teaching librarians, improve our own practice. We hope they will do the same for you.

Lori Townsend and Silvia Lin Hanick
Series Editors

PART I
OVERVIEW

CHAPTER 1
INFORMATION LITERACY: WHAT IT IS AND WHY IT MATTERS

Don't skip this chapter thinking it's just boring introductory material! It's not! It explains what information literacy skills are, what specific skills you can expect to learn from this book, and why those skills are particularly important for scientists and engineers. You'll also learn how the storage of information in **sources**, the retrieval of those sources from **databases**, and the **access** to sources that libraries provide are all critical to the advancement of science and engineering.

Information Literacy Is a Set of Skills

Information is more abundant and complex than ever before and we use it constantly for school, for work, and for all the decision-making we do in life. To navigate this complexity, it helps to have strong information literacy skills. But what is **information literacy**? The Association of College and Research Libraries (ACRL) defines it as follows:

> Information literacy is the set of integrated abilities encompassing the reflective discovery of information, the understanding of how information is produced and valued, and the use of information in creating new knowledge and participating ethically in communities of learning.

To provide science and engineering students (like you) with these abilities, we need to begin with some very basic skills that'll help you understand, find, and use **information sources**. After all, scientists and engineers spend a lot of time interacting with information found in sources like books, journal articles, conference papers, and websites. Scientists and engineers must regularly refer to these sources to educate themselves, to keep up to date on the latest best practices, to make informed decisions, and to build upon existing research.

Because information is delivered in sources, it makes sense to focus on finding, evaluating, and using sources as the foundation of your information literacy skill set. After gaining this foundation, you'll be able to reflect more deeply on how information is produced, valued, and used, and assess the strengths and weakness of those processes. This foundation will also free up bandwidth for you to critically assess the information presented in sources. It's much easier to process, interpret, and judge information you've read in a source if you first understand what the source is, why it exists, and the typical qualities it has.

So now, let's review the basic skills you'll learn in this book. These skills directly relate to understanding sources, searching databases to find sources, and citing sources:

Skills for understanding sources:

- Distinguish sources from one another and understand their purpose
- Identify sources in both their print and online formats
- Recognize the qualities of a source and whether those qualities suit your needs
- Know how to find sources, obtain the full text, as well as store and organize your sources

Skills for searching databases to find sources:

- Understand different types of databases and how they work
- Know how to find databases and choose the best to search
- Search databases effectively for sources

Skills for citing sources:

- Understand why to cite sources
- Know when and how to cite sources

You'll need to draw upon these skills while you're in college and if you pursue a graduate degree, as well as throughout your career. In fact, many of the skills taught in this book are geared toward helping you with research challenges you'll encounter during your career. But this book also contains tips to help you do well on the types of research assignments you'll work on in college and graduate courses.

Gaining a good set of information literacy skills will help you conduct research while you're in college or graduate school—as well as throughout your career—but there are two meanings for the word "research" so let's be clear what type of research we're referring to. The process of searching and finding information is called **conducting research**, and this book focuses on teaching the skills you'll need to do that.

However, when you carry out experiments and gather data, that's also called **conducting research**. So, research can take place in two different settings which we can differentiate as the **library** or the **lab**. For both, the goal is to find something out—but library research usually consists of gathering and using pre-existing information, while lab research consists of conducting studies and experiments to learn something brand new.

Who Should Learn the Information Literacy Skills Explained in This Book?

The skills and concepts explained in this book are essential for both undergraduate and graduate students in science and engineering, as well as for students in some of the social sciences. If you are studying any of the disciplines (or associated subdisciplines) found in Table 1.1 then you'll find this book helpful.

Table 1.1

Physical Sciences	*Life Sciences*	*Applied Sciences*	*Social Sciences*
Atmospheric science Astronomy Chemistry Climatology Earth science Geology Meteorology Physics	Biochemistry Biology Botany Ecology Environmental science Forestry Microbiology Paleontology Zoology	Biomedical engineering Chemical engineering Civil engineering Computer science Electrical engineering Health sciences Mechanical engineering Medicine Nursing	Education Physical anthropology Physical geography Psychology Sociology

Being Information Literate is Easy in Some Ways—But Hard in Other Ways

You probably find it easy to search Google for quick answers to simple questions like "When did the Grand Canyon become a national park?" Also, you're probably adept at searching Google for basic, introductory information. For example, you could easily find websites that provide examples of the types of animals that lived during the Jurassic period.

In the old days, before the internet existed, it was not nearly as easy to get quick answers to questions like these. So, in some ways, finding information has become a lot easier over time. However, in other ways, finding information has become harder. That's because with each passing day more information comes into existence, making it tougher to sift through and find what you need. Plus, nowadays there are more *types* of sources than ever before, and many of those sources exist in multiple formats. Whereas in the past sources only existed in **print**—meaning their paper-based, physical format—many sources are now digitized and available **online**, *in addition* to existing in print (see Figure 1.1). It's challenging to know where and how to look for information when it has become so much more abundant and complex.

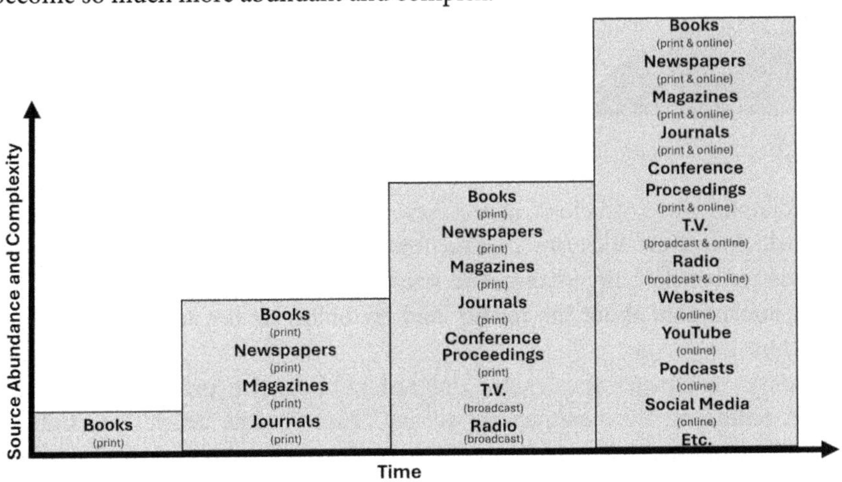

Figure 1.1

The sheer number and complexity of sources can be especially challenging when your information needs are advanced and/or high stakes, as they tend to be for scientists and engineers. Let's look at an example. Suppose you work for the United States National Park Service and need to prepare a comprehensive report on strategies for mitigating erosion caused by hikers and campers in Grand Canyon National Park, including the pros and cons of different strategies, the costs, etc. It wouldn't be possible to gather this information with a few quick Google searches. You'd need more sophisticated information literacy skills to research this properly. Specifically, you'd need to know what types of sources are likely to have the information you need, which databases to search to find those sources, how to conduct a thorough search, and how to choose the most useful and relevant sources from what you found.

Furthermore, scientists and engineers usually need to find *high-quality* information for the sort of work they do. Yet, it has become more difficult to identify and separate high-quality information from low-quality information when there is so much information out there, *and* when you encounter most of it digitally on the internet. Digital information all looks a bit uniform and "samey" when it's beaming out from your computer monitor, tablet, or phone, making it hard to determine what *type* of source you're looking at, let alone whether it's high or low quality. However, back when information was only published in physical, print-based sources, it was much easier to recognize and distinguish different source types. Plus, the way a source looked—including its size, paper-type, amount of color, number of images, etc.—offered lots of clues about the source's purpose, audience, quality, and credibility.

Figure 1.2

For example, when you look at the newspaper covers shown in Figure 1.2, can you tell which is more likely to contain informative, reliable articles versus gossipy, sensationalized articles? The images, font sizes, the tone of the headlines, etc., all supply you with information about the quality and credibility of the articles printed on the pages within.

Nowadays, newspaper articles are not just published in print form, they're also published online on the newspaper's website. These online articles are indexed by Google, which means they can appear in your Google results, and that's where most students encounter them. But when you click on a Google result that leads to an online

newspaper article, you'll just see the text of the article *without* that super-informative newspaper cover, leaving you with fewer clues about the quality and credibility of the article. Actually, you may not even realize you're looking at a newspaper article, which puts you at an additional disadvantage. Information is packaged in different types of sources because those sources serve different purposes. If you can't distinguish one type of source from another, then it's harder to assess whether the source you're viewing has content appropriate for your needs.

Conducting a search that is *thorough* has also become harder, again because there is so much information out there. Why must scientists and engineers conduct thorough searches? One reason is to ensure they are not embarking on a research project that someone else has already done. What a waste of time, money, and effort that would be, right? Thorough searching is also necessary when you need to investigate best practices or come up with new policies. For example, the research project described earlier—about mitigating erosion caused by hikers and campers at the Grand Canyon—would require a *thorough* search of information before you could assimilate and put forward best practices or policy recommendations.

You Might Need to Improve Your Information Literacy Skills More Than You Think

Many students are confident they already have excellent information literacy skills, but most are overestimating their abilities. It's not just students, *everyone* overestimates their information literacy skills. It's understandable why this happens. For many of our information needs, we're able to easily find answers using Google. This leads us to assume we must be pretty good at finding, evaluating, and using information. But the study below found something different:

Breakstone, J., et al. (2019). *Students' Civic Online Reasoning: A National Portrait.* Stanford History Education Group and Gibson Consulting. https://purl.stanford.edu/gf151tb4868

The authors of that study gave students a series of tests to assess their ability to recognize poor-quality, false, and/or biased information on the internet. This quote from the study presents some of the results:

> **Fifty-two percent** of students believed a grainy video claiming to show ballot stuffing in the 2016 Democratic primaries (the video was actually shot in Russia) constituted "strong evidence" of voter fraud in the U.S. Among more than 3,000 responses, only three students tracked down the source of the video, even though a quick search turns up a variety of articles exposing the ruse.
>
> **Two-thirds** of students couldn't tell the difference between news stories and ads (set off by the words "Sponsored Content") on Slate's homepage.
>
> **Ninety-six** percent of students did not consider why ties between a climate change website and the fossil fuel industry might lessen that website's credibility. Instead of investigating who was behind the site, students focused on superficial markers of credibility: the site's aesthetics, its top-level domain, or how it portrayed itself on the About page.
>
> Nearly all students floundered. Ninety percent received no credit on four of six tasks.

So, lots of students in the study thought they were viewing relevant, credible, sources, but instead they were viewing false information, advertisements, or biased information—and they had no idea. Imagine graduating from college, landing a job, and then compiling an important report for your boss using internet sources that are of very low quality! That's not a good way to excel at (or keep) your job.

Students' tendency to overestimate their information literacy skills has been documented in other studies as well, and this tendency is usually attributed to the **Dunning–Kruger effect**. What's that? It's when people think they're good at something when they're not. Take a look at the title of the study where this psychological phenomenon was first documented:

> Kruger, J. & Dunning, D. (1999). Unskilled and unaware of it: how difficulties in recognizing one's own incompetence lead to inflated self-assessments. *Journal of Personality and Social Psychology*, *77*(6): 1121–1134.

Below are a few examples of studies that investigated the Dunning–Kruger effect in relation to college students and their information literacy skills:

> Gustavson, A. & Nall, H. C. (2011). Freshman overconfidence and library research skills: a troubling relationship? *College and Undergraduate Libraries*, *18*(4): 291–306.
>
> Mahmood, K. (2016). Do people overestimate their information literacy skills? A systematic review of empirical evidence on the Dunning–Kruger effect. *Communications in Information Literacy*, *10*(2): 199–213.

Molteni, V. E. & Chan, E. K. (2015). Student confidence/overconfidence in the research process. *The Journal of Academic Librarianship*, *41*(1): 2–8.

Jackson, C. (2013). Confidence as an indicator of research students' abilities in information literacy: a mismatch. *Journal of Information Literacy*, *7*(2): 149–152.

The titles of these articles tell the story pretty well—there's a mismatch in students' assessment of their skills and their *actual* skills. So, what does this mean for you? It means that information literacy skills will be *harder* for you to learn if you think you're already good at them. The best way to start improving your information literacy skills is to set aside any assumptions you have about your ability to conduct research, and just start fresh!

Adjust Your Expectations about the Time and Effort That Research Requires

The ease of searching Google has led everyone to expect that research should *always* be quick and easy. But that expectation is not only unrealistic, it leads to serious problems. In fact, a study conducted a few years ago by a research group called **Project Information Literacy** investigated how employers view the research skills of newly hired college graduates:

Head, A. J. (2012). *Learning Curve: How College Graduates Solve Information Problems Once They Join the Workplace.* Project Information Literacy Research Institute. https://projectinfolit.org/publications/workplace-study/

The study found that employers were unhappy "that college hires rarely conducted the thorough research required of them in the workplace." Employers said their new hires "were prone to deliver the quickest answer they could find using a search engine, entering a few keywords, and scanning the first couple of pages of results." In other words, these employers were unhappy that newly hired college graduates approached research like it was something to get over and done with as quickly as possible, instead of taking the time to do it right.

So, as you make your way through college and/or graduate school and then embark on your career, think of research as a process of discovery that deserves your time and effort. Slow down, be patient, and expect that research will sometimes be time-consuming. It's usually the time-consuming research that's the most rewarding and interesting, so savor it!

How Does Information Literacy Relate to Science and Engineering?

New discoveries and inventions don't just emerge from nothing; rather, they come about when researchers build upon facts and information originating from earlier discoveries. Take, for example, this journal article:

Craig, L. S., et al. (2008). Stream restoration strategies for reducing river nitrogen loads. *Frontiers in Ecology and the Environment*, 6(10): 529–538.

That article describes strategies for reducing the amounts of environmentally damaging nitrogen (originating from agricultural fertilizers) that drains into streams and rivers. It wouldn't have made sense to write an article about how to *reduce* nitrogen pollution unless there had already been research establishing that (1) nitrogen is a pollutant, and (2) it causes enough damage that we need to do something about it. So, that article builds upon and extends findings that were established in previously published research.

Researchers are constantly drawing upon information from previously published studies to provide background, context, and direction for new studies. That's how we make progress—by accumulating facts and then expanding research efforts into new areas. But none of this can happen unless researchers are able to share their progress with each other across space and time. In fact, we need a way for *any* researcher to be able to search for and refer to the work that *any other* researcher has done. In other words, we need a documentation and retrieval system that meets these four conditions:

1. Research discoveries must be publicly recorded and stored in sources.

To be of future use to others, research discoveries must be publicly recorded and stored. **Publishers** fulfill this role by recording and storing research results in **sources** that they make available in print or online (or both)—usually for a fee. In science and engineering, new research discoveries are recorded in two main types of sources: journal articles and conference papers. But information about research discoveries also ends up in books, newspaper articles, and other types of sources. You'll learn all about these sources in Chapters 3–13, but for now just note that one of the main functions of sources is to document information for posterity. Without sources, we'd have no record of what we already know.

2. Sources must be findable.

There's no point in recording and storing information in sources unless that information can be found and used. Nowadays, we almost exclusively use **databases** for finding sources. For example, if you were looking for journal articles on a particular topic, you could search a free database like Google Scholar to find them. Or, if you were using your campus library (which, by the way, is called an **academic library**), you could search the databases your library offers on its website which might include Scopus, GeoRef, Compendex, or Web of Science—to name just a few examples.

3. Sources must be accessible.

Even if you can find appropriate sources to consult, they won't be of use to you unless you can access them and read them in their entirety. When you search a database and find a promising source, there's usually a link to the full text of the source somewhere in the database record. When you click that link, whether you can actually view the full text of the source depends on whether the source happens to be free, or whether you can afford to pay a fee to the publisher to view it, or whether you are affiliated with a library that has purchased or subscribed to it. That's right, your campus library pays fees to provide you with access to sources like journal articles, newspaper articles, books, etc., so that you don't have to pay for them yourself. As most sources are not free or cheap, libraries buy and collect them, thus solving the accessibility problem. Just imagine you're conducting research on how to design concrete buildings to withstand earthquakes, but you can't afford to buy and consult this expensive book:

Moehle, J. (2015). *Seismic Design of Reinforced Concrete Buildings*. McGraw Hill Education.

Then, you type the title of that book into the search box on your campus library's website to see if your library has it, and it does! In Figure 1.3 you see the record for it including a link to access the full text online, right from your computer.

Figure 1.3

Even when your campus library doesn't own a source you need, it'll provide a service known as **document delivery** or **interlibrary loan** to help you get the source from another library that has it. So, libraries play an important role in your ability to access sources, not just by purchasing and subscribing to various sources but also by providing services like interlibrary loan/document delivery.

4. Researchers must have appropriate skills for finding, accessing, and using sources.

Sources don't find and use themselves—researchers like *you* must have the appropriate skills to locate and use them. One of those skills is knowing what types of sources are likely to have the information you need. You'll also need to know what databases to use to find those types of sources. You'll need to know how to search those databases with keywords, and you'll need even more sophisticated search skills for times when your search must be thorough. You'll also need to know how to interpret your database results and how to manipulate and refine your searches to retrieve better results. You'll need to be able to choose the most relevant sources from your results and access the full text of those sources. Last but not least, you'll need to know how to process the information you gathered from the sources you found and incorporate those sources into your writing, citing them appropriately.

See how it all fits together? Because researchers need to communicate information to each other over space and time, we've developed a system to make that happen. Publishers record and store information in sources. Those sources are made findable using databases. Libraries pay publishers to make both sources and databases accessible. But none of it matters without the involvement of you and your information literacy skills, so let's start improving them!

To sum up, what should science and engineering students know about information literacy?

1. Scientists and engineers are in disciplines that grow and expand only by building on previously established facts and discoveries that have been published in sources. That means they need solid information literacy skills for finding, evaluating, and using sources. This book will help you learn those skills!
2. Over time, information literacy skills have become more challenging to learn as information has expanded in abundance and formats.
3. Most people—including students and even professionals—don't think they need to improve their information literacy skills, and this impedes their ability to get better at finding, evaluating, and using sources.
4. There's a tendency to expect research to be quick and easy, but this is not always the case. When you need to conduct thorough and comprehensive research on a topic, then expect research to take longer.

What's Next?

You just finished **Part I: Overview**, which consists of just this one chapter. Hopefully this chapter helped you understand what information literacy is and why it's so important. Now you'll move on to **Part II: Sources**. In this set of chapters, you'll learn about different types of sources, their functions, how to distinguish them from each other, and where to find them. The first chapter in Part II explains how information comes into existence and ends up in sources.

PART II
SOURCES: THEIR FUNCTIONS, DISTINCTIONS, AND WHERE TO FIND THEM

CHAPTER 2
INFORMATION IS DOCUMENTED IN SOURCES

This chapter explains how information is "born" and then recorded in sources. That process involves publishers, and you'll learn about those that specialize in science and engineering content. You'll also learn that different types of sources have different purposes and qualities. That means when you have an information need, not just any source will resolve it. Rather, you can save time and effort by seeking answers in the type of source that's best matched to your information need. To do that, you must be able to distinguish different types of sources from one another, which you'll learn to do in the chapters that follow this one.

New Information Is Generated in a Variety of Different Ways

New information is constantly coming into existence, and there are various ways this happens. Solving everyday problems creates new information, such as developing a tastier recipe for gluten-free brownies. However, that information may never end up in a published source unless it gets recorded and shared in a cookbook or a cooking blog.

Writing an email is also an act of information creation. For example, you might email your parents with information about how you're doing in college or graduate school. But your parents are probably the only people interested in that information, so there's no need to record it in a publicly accessible source and share it with others.

New information also results from research conducted by private companies, such as engineering companies or pharmaceutical companies. This information is usually documented and stored within the company, but not shared outside the company. Protecting this **proprietary information** (also known as **trade secrets**) is what allows companies to gain a business advantage over their competitors.

As these examples show, newly generated information is not always publicly shared. So now let's focus on when and how new information *is* publicly shared with others through the act of recording it in sources.

Newsworthy Events Are Recorded in Sources

Brand-new information in the form of newsworthy events is recorded in sources like newspapers. What's a newsworthy event? When changes happen in the world, and those changes are of interest to lots of people, then that's a **newsworthy event**. Examples of newsworthy events include election results, natural disasters, corporate scandals, scientific breakthroughs, etc. A newsworthy event becomes a published **source** when a journalist writes it up as an article and publishes it in a **newspaper** (like *The New York Times*), or in a **news magazine** (like *Time* or *Newsweek*), or on a **news website** (like HuffPost, MSN.com, Politico, or The Hill).

Scientific and Technological Discoveries Are Recorded in Sources

New information is also generated through scientific and technological discoveries. For example, biomedical researchers might design an experiment to determine which type of blood-thinning drug is most effective for cancer patients who are at risk of developing blood clots, as they do in this journal article:

> Lee, A. Y., et al. (2015). Tinzaparin vs. warfarin for treatment of acute venous thromboembolism in patients with active cancer: a randomized clinical trial. *JAMA, 314*(7): 677–686.

Or engineers might figure out how to extend the flight duration of a drone by incorporating an automatic battery replacement system, as they do in this conference paper:

> Fujii, K., et al. (2013). Endless flyer: a continuous flying drone with automatic battery replacement. *2013 IEEE 10th International Conference on Ubiquitous Intelligence and Computing and 2013 IEEE 10th International Conference on Autonomic and Trusted Computing* (pp. 216–223). IEEE.

When scientists and engineers conduct research to learn something new, they publish their findings in sources like **journal articles** and **conference papers** to share the information with others. In Chapters 6–9, you'll learn more about the important role of journal articles and conference papers in disseminating research.

Previously Published Information Is Taken and Repurposed in Different Sources

You've just seen a few ways that newly generated information is recorded and shared in sources. Over time, information from these sources might be analyzed, synthesized, or summarized—and then presented in different types of sources. For example, consider this newspaper article from *The New York Times*, in which a journalist reports on Boeing 737 airline crashes:

> Gelles, D. (2019). Boeing 737 Max: what's happened after the 2 deadly crashes. *New York Times*. https://www.nytimes.com/interactive/2019/business/boeing-737-crashes.html

Over time, a historian might find and use that article as a source of information for a book on the history of airline disasters. Or, the lessons learned from Boeing 737 malfunctions might end up in a textbook on aeronautical engineering or in a handbook on aviation risk management.

Here's another example. Research establishing that nitrogen pollution is harmful to streams and rivers was first recorded and published in journal articles. Over time, researchers would conduct more studies and publish more journal articles confirming the negative effects of nitrogen pollution. As evidence for the detrimental effects of nitrogen pollution grew, it would gain more traction and acceptance, until finally it would be considered an established fact. Facts get incorporated into all kinds of different sources so that more people have a chance to learn about them. For example, you can read the Wikipedia entry on "Water pollution" and find out about nitrogen as a water pollutant.

Facts about nitrogen pollution might also end up in a college or graduate textbook on freshwater ecology. Sources like textbooks serve the particular needs of students. Students can learn more easily from a textbook because it provides an overview of accumulated facts on a topic. Imagine how much harder it would be to read the hundreds of dense, technical journal articles those facts were originally derived from!

When information is taken from one source and used in another source, it's usually altered in some way. For example, information taken from journal articles will be summarized and simplified when it's used to compile a book that presents a comprehensive overview of a topic. Consider, for example, this book:

> Beckman, D. (2012). *Marine Environmental Biology and Conservation*. Jones and Bartlett Publishers.

Most of the information in that book originated from research studies published in journal articles. As facts accumulated from all those research studies, the author of this book gathered them together and wrote a book explaining the basic principles of marine environmental biology and conservation.

Data offer another example of information that gets repurposed and presented in different sources. Consider that daily weather records, pollen counts, tides, and moon phases will be incorporated into sources like weather websites and almanacs. Or data might end up in a book, like the one below, which uses pollen and temperature records to predict the effects of climate change on allergens:

> Beggs, P. J. (Ed.) (2016). *Impacts of Climate Change on Allergens and Allergic Diseases.* Cambridge University Press.

In another example, the journal article below uses census data that were originally gathered by a government agency to look at relationships between suicide, socioeconomic status, and isolation:

> Näher, A. F., et al. (2020). Associations of suicide rates with socioeconomic status and social isolation: findings from longitudinal register and census data. *Frontiers in Psychiatry*, DOI: 10.3389/fpsyt.2019.00898

So, these examples show how information gets taken from one source and then presented in another source—often to serve a different purpose. This sort of activity is happening constantly because humans are insatiable information processors!

Two More Important Ways That Information Is Repurposed

Previously published information is also compiled, analyzed, summarized, and repurposed in two important types of sources called **review articles** and **systematic review articles**. A **review article** is a special type of journal article where the authors gather all the studies that have been conducted in a specific area of research and then write up an overview of everything we know about that area of research.

A **systematic review article** is similar, but a bit different. In this type of article, the authors gather research studies that have investigated a particular research question. They compile the data and results from those studies and then write up a systematic review article that discusses whether the studies are all producing evidence that converges on the same answer to the research question. If so, the systematic review article would show a high level of confidence that we have the right answer to that research question. Here's an example of a systematic review article that examines what the collective evidence says about managing stress:

> Hannigan, B., et al. (2004). Stress and stress management in clinical psychology: findings from a systematic review. *Journal of Mental Health, 13*(3): 235–245.

Because these two types of articles are so useful, you'll learn more about them in Chapter 8.

Information Ends Up in Sources through the Act of Publishing

Information is stored and publicly shared in sources through the act of **publishing.** Typical examples of published sources include newspaper articles, magazine articles, journal articles, and books. Note that anything posted to the internet, such as a blog entry or a Wikipedia article, is also considered to be a published source. So, anyone can publish information simply by posting it to the internet and making it public. However, publishing—in its more traditional sense—is an enterprise that takes place through a publisher. A **publisher** is a company or organization that accepts manuscripts from authors, provides editing services, and manages the layout, printing, webhosting, advertising, and distribution of the source in its finished form.

There are different types of publishers and some specialize in producing certain types of sources. For example, The New York Times Company is a newspaper publisher. Plus there are magazine publishers, such as Conde Nast, and book publishers, such as Penguin Books.

There are also publishers that specialize in producing sources for scientists and engineers (as well as other professionals), including:

1. **Commercial Publishers**—Commercial publishers are companies that operate for a profit. Most publish several different types of sources including books, journals, and conference proceedings. Examples of commercial publishers that provide science and engineering content include Springer, Elsevier, Wiley, and SAGE.
2. **Society Publishers**—Society publishers are associated with professional organizations or societies, and they usually produce several different types of sources, just like commercial publishers. Society publishers that provide science and engineering content include the Institute of Electrical and Electronic Engineers (IEEE), the Association for Computing Machinery (ACM), and the American Chemical Society (ACS)—just to name a few.
3. **University Publishers** (also known as **University Presses**)—University presses are publishing operations associated with academic institutions and they're mostly known for publishing books. A few examples include Cambridge University Press, Harvard University Press, and Princeton University Press.
4. **Open-Access Publishers**—Most publishers recoup their costs and generate profits by charging fees—such as subscription fees—to anyone who wants to read their published content. But open-access publishers operate under a different economic model in which a payment is made up front to get the manuscript published. This payment is called a **publication fee** or an **article processing charge** and it's often paid out of the author's research grant-funding or with funds made available by the author's employer. Because these fees cover the cost of publication there's no need to make the reader pay anything. In other words, anyone wanting to read the publication can do so for free without needing to pay subscription fees or access charges. Most open-access publishers specialize in producing journals. Examples include *Public Library of Science (PLoS)* and *BioMed Central (BMC)*. You can learn more about open-access publishers and the open-access publishing movement in Appendix C.

Just so you know, there are also **predatory publishers** that produce **predatory journals**, which are not legitimate journals. These publishers prey upon faculty at academic institutions who are anxious to increase the number of articles they've authored. Why are faculty anxious about this? Most academic employers regard publishing as a measure of a faculty member's productivity, so publishing very few articles is equivalent to being unproductive and could lead to losing your job. If you've heard the saying "Publish or perish!" that's what it refers to.

Predatory publishers send emails to faculty at academic institutions encouraging them to submit article manuscripts. The predatory publisher promises the author quick acceptance and publication of their manuscript. The author must pay a publication fee to get their article published, just as they would with an open-access publisher, but that's where the similarity ends. Predatory publishers are really just exploiting the open-access publishing model by collecting fees from authors who are anxious to increase their number of publications. Predatory publishers don't bother to ensure that the articles they publish are high quality or even valid, in contrast with legitimate publishers who employ teams of editors and utilize a peer-review process to ensure quality. (You'll learn all about the peer-review process in Chapter 14.)

If you're interested in learning more about predatory publishers, then here are two good sources:

> Linacre, S. (2022). *The Predator Effect: Understanding the Past, Present and Future of Deceptive Academic Journals.* Against the Grain Media. https://doi.org/10.3998/mpub.12739277

> Grudniewicz, A., et al. (2019). Predatory journals: no definition, no defence. *Nature, 576*(7786): 210–212.

But don't worry too much about predatory publishers right now because you'll learn how to identify them in Chapter 6. Also, as you gain more experience, you'll start to recognize the reputable and upstanding publishers that serve your discipline.

Different Sources Serve Different Purposes

Information is published in different types of sources for a *reason*. It's not just random that we have books, journals, newspapers, etc. These different source types have specific purposes that fulfill different information needs. Knowing this can save you time because you can focus on looking for the *type* of source that's most likely to have the information you need. For example, if you're interested in keeping up with political news, then newspapers and news websites are ideal, but you'd be wasting your time if you tried to find the latest political news in a journal article or book.

Or, if you need a broad, comprehensive overview of a topic, such as how to manage civil engineering projects, you'd be better off reaching for a book, such as:

> Twort, A. & Rees, G. (2020). *Civil Engineering Project Management.* CRC Press.

But if you're gathering information about a specific research topic, such as the effects of invasive cheatgrass on fire risk in the Western United States, then your best bet will be to look for journal articles, like this one:

> Bradley, B. A., et al. (2018). Cheatgrass (*Bromus tectorum*) distribution in the intermountain Western United States and its relationship to fire frequency, seasonality, and ignitions. *Biological Invasions*, *20*(6): 1493–1506.

In fact, there are lots of types of sources, each with its own particular purpose and qualities that serve specific information needs. Here are some examples of source types, just to give you an idea of the variety that's out there:

- Books
- Magazine articles
- Newspaper articles
- Journal articles
- Conference papers
- Websites
- Data
- Patents
- Standards
- Codes
- Protocols
- Technical reports
- Technical papers
- White papers
- Maps
- Photographs
- Videos
- Sound recordings
- Archival materials

We won't investigate all these source types within the pages of this book—rather, we'll just focus on the sources in the left column of the list above, as those are the most heavily used by scientists and engineers.

Distinguishing Sources from One Another Is a Key Skill

Even if you know what *type* of source is likely to have the information you need, this won't save you time and effort unless you can *recognize* and *identify* different source types, especially in their online form, which is where you'll most likely encounter them. However, as you learned in Chapter 1, sources in their online form tend to look alike and can be hard to distinguish from one another.

Also, many people find it hard to distinguish source types in their Google results. Suppose you were searching Google for information about climate change modeling software for use by insurance companies, and you came up with the results shown in Figure 2.1. Would *you* be able to identify what types of sources your results represent? Because experienced researchers *can*.

Experienced researchers pick up on clues in their Google results to identify the type of source the result represents. For example, most researchers would know that the

Figure 2.1

second result (shown above) is for a magazine article because it's coming from a well-known magazine called *Smithsonian Magazine*. The third result is coming from *The Wall Street Journal*, which is a financial newspaper that most researchers are already aware of.

Meanwhile, the last result is coming from a reputable science and engineering journal publisher called "Springer," and being delivered on their platform *SpringerOpen*. Plus, within the description of this result, you see it's been "Cited by 24." This is a metric or score that's associated with journal articles. Together these two pieces of information (the publisher, and the existence of a citation metric) make it obvious to experienced researchers that this result leads to a journal article. (You'll learn more about citation metrics Appendix B.) Because that first result isn't coming from any recognizable traditional publication, then chances are high it's a website.

So, experienced researchers rely on their accumulated knowledge of publishers and sources to quickly identify source types within their Google results, and this helps them make informed choices about what to click on. For example, if you need scholarly, authoritative information on the topic of climate change modeling software, the last Google result in the image shown above (the journal article) would be the best to click on, while the other results would probably be less useful to examine. But, if you were looking for simple, non-technical information on this topic, the journal article might be the *worst* result to click on, because journal articles can be chock-full of technical vocabulary! So, knowing the types of sources your Google results represent can help you quickly zero in on information that has the exact qualities you need.

However, for less-experienced researchers, it's hard to distinguish different source types in a list of Google results. Plus, even after you click on a result and look it over, it might not be clear what type of source you're viewing. This puts you at a real disadvantage because you don't know anything about the purpose of the source you're viewing or the qualities you can expect it to have, such as whether it's **scholarly** or **primary** or **credible**, etc. You might *think* you're already good at research, but unless you can recognize source types and incorporate your understanding of them into your research process, you don't know what you're missing! In fact, take a look at the journal article in Figure 2.2, which explores the problems students have identifying source types (or "genres," as this author calls them) in an online environment.

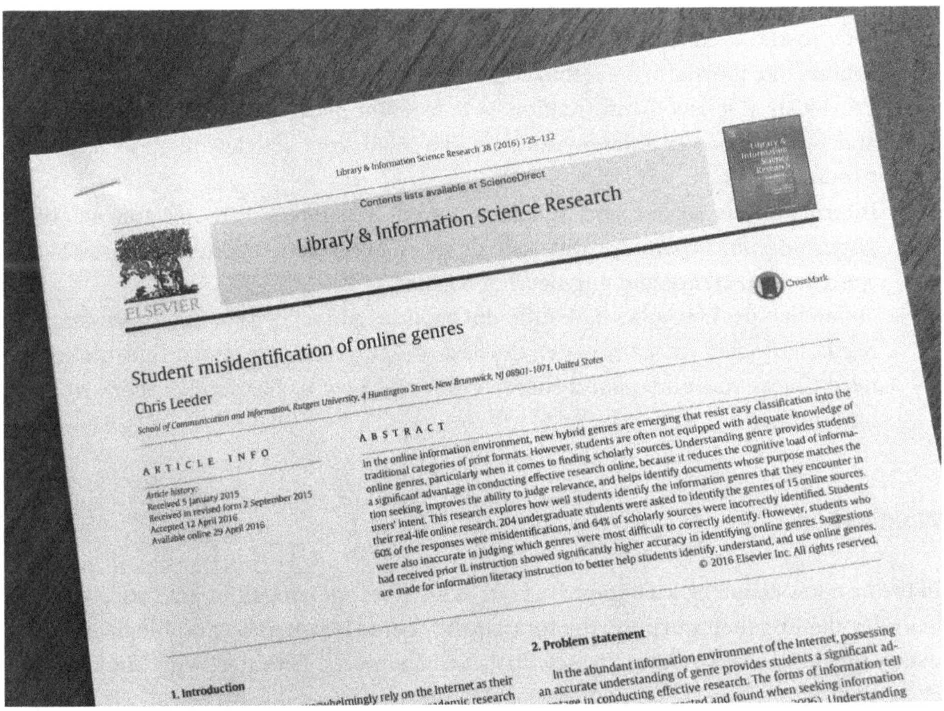

Figure 2.2

Leeder, C. (2016). Student misidentification of online genres. *Library and Information Science Research*, *38*(2): 125–132.

The author of this article explains why it's useful for students, or anyone really, to able to identify source types (or genres) in their online form:

> In the abundant information environment of the Internet, possessing an accurate understanding of genre provides students a significant advantage in conducting effective research … Understanding genre provides context by incorporating an understanding of a document's purpose or function, and allows for finding documents whose purpose matches the user's intent and helps them make judgments regarding the document's relevance to their information need (Crowston et al., 2010, Rosso, 2008).

So, it's pretty useful to be able to recognize different types of sources and understand the purpose and qualities of those different sources. Mastering this skill will make you much more efficient and effective at conducting research.

To sum up, what should science and engineering students know about how information is documented in sources?

1. New scientific and technological discoveries are documented and shared in sources like journal articles and conference papers.
2. Previously published information is constantly being gathered, synthesized, and repurposed in different sources like Wikipedia articles, textbooks, and review articles.
3. Information is stored and publicly shared in sources through the act of publishing, and there are different types of publishers, including those that specialize in science and engineering content.
4. Different types of sources have different qualities and serve different information needs. However, you can't match the best source to your particular information need unless you understand source types and how to distinguish them from one another.

What's Next?

In the next few chapters, we'll improve your skills at recognizing different sources as well as understanding their purpose, the formats they come in, where to find them, and what makes them especially relevant to scientists and engineers. Let's start with **books**!

CHAPTER 3
BOOKS

In this chapter you'll learn about the types of books used by scientists and engineers. This chapter will also explain where to search and find books, as well as what to do if you can't find a book you need. You'll also learn how to distinguish online books from other sources.

An Overview of Books

Of all the different types of sources we use today, books are the oldest. The first books came into existence long before other types of sources, and their abundance greatly increased after the invention of the printing press (first in China, then later in Europe in 1450). Over time, we've accumulated about 130 million books and this number grows with each passing year. But let's focus on books that explore science and engineering topics. Typically, the objective of these books is to explain a topic comprehensively in order to inform and educate.

Science and engineering books can be categorized into four main types: popular, scholarly, textbook, and reference. However, the lines between these types can sometimes be a bit blurry—for example, a book might fall somewhere in between popular and scholarly, or a scholarly book might also serve as a textbook.

Popular Books

Popular science and engineering books explain ideas, concepts, or history to the non-expert with the aim to inform, but in a way that's captivating and entertaining. Here are some examples:

Mukherjee, S. (2010). *The Emperor of All Maladies: A Biography of Cancer*. Simon and Schuster.

Harari, Y. N. (2019). *Sapiens: A Brief History of Humankind*. Harper.

Ellenberg, J. (2015). *How Not to be Wrong: The Power of Mathematical Thinking*. Penguin.

Scholarly Books

Scholarly books, also known as **academic books** or **monographs**, comprehensively examine a research topic with the goal to inform. Unlike popular books, scholarly books are not written for the general public to read and enjoy. Instead, their main audience is other scholars and experts, or anyone wanting to develop expertise on a topic, such as undergraduate or graduate students. Here are a few examples of scholarly books relevant to scientists:

Ellis, G. & Solms, M. (2017). *Beyond Evolutionary Psychology: How and Why Neuropsychological Modules Arise*. Cambridge University Press.

Roni, P. & Beechie, T. (Eds.) (2012). *Stream and Watershed Restoration: A Guide to Restoring Riverine Processes and Habitats*. John Wiley & Sons.

Heikkila, W. (2011). *Earth's Magnetosphere: Formed by the Low-Latitude Boundary Layer*. Elsevier.

Textbooks

Like scholarly books, textbooks comprehensively examine a particular topic but with the goal of educating the reader. To that end, they often include learning objectives, study questions, vocabulary words in bold, and other features designed to facilitate learning. Many college courses require you to buy and read textbooks. Here are some examples:

Raven, P. H., et al. (2019). *Biology*. McGraw Hill.

Pond, R. J. & Rankinen, J. L. (2013). *Introduction to Engineering Technology*. Pearson.

Plummer, C., et al. (2019). *Physical Geology*. McGraw Hill.

Reference Books

Reference books are a bit different because their purpose is not to explain a topic. Instead, they contain information like facts, definitions, measurements, or procedures that you'd look up and **refer** to—which is why they're called **reference** books. **Dictionaries** and **encyclopedias** are the most familiar examples. However, nowadays it's common to use free, online dictionaries as well as sources like Wikipedia rather than using physical versions of dictionaries and encyclopedias. In fact, an enormous amount of reference information is now quick and easy to find using Google. But not everything! There's still a lot of information that's not conveniently available on the internet, such as content found in dictionaries and encyclopedias specialized for scientists and engineers. Here are some examples:

Lackie, J. M. (Ed.) (2007). *The Dictionary of Cell and Molecular Biology*. Academic Press.

Chatti, S., et al. (Eds.) (2019). *CIRP Encyclopedia of Production Engineering*. Springer.

Selley, R. C., et al. (2004). *Encyclopedia of Geology*. Academic Press.

Scientists and engineers use other types of reference books as well, such as **handbooks** and **manuals**. These typically contain facts or instructions, or some combination of both. Here are some examples:

Lide, D. R. (2010). *CRC Handbook of Chemistry and Physics*. CRC Press.

VanderPlas, J. (2016). *Python Data Science Handbook: Essential Tools for Working with Data*. O'Reilly Media, Inc.

Thyer, B. A. (2009). *The Handbook of Social Work Research Methods*. SAGE Publications.

The *CRC Handbook of Chemistry and Physics* provides the properties of elements and fluids, as well as physical constants of organic compounds. The *Python Data Science Handbook* provides instructions on how to use Python to manipulate and visualize data. And *The Handbook of Social Work Research Methods* outlines different ways to conduct research using sampling and statistics.

There are many other types of reference books, such as **atlases**, **gazetteers**, and **biographical dictionaries**, to name a few. Plus, some reference books are designed to fulfill the specific needs of experts in certain disciplines. For example, botanists might use a type of reference book called a **field guide** to identify plants.

Basically, if you're looking for discrete bits of information such as statistics, measurements, properties, procedures, standards, etc., and you can't quickly find what you need on the internet—or, you can't find what you need on a website that you trust to be reliable—then you should stop and ask yourself: "Would there be a reference book that has this information?" Better yet, ask a librarian at your campus library this question. There's no point wasting time searching the internet for information that simply isn't there. Instead, find the reference book that has it.

Books Tend to Be Less Current than Other Sources

Because books are thorough and lengthy, they take longer to write. That means they may contain outdated information before they're published and ready for sale. As many science and engineering fields advance rapidly, books published in these fields can quickly become outdated. When this happens, book authors will sometimes update their book and publish a new **edition**. So, just be aware that if you're looking for a book

on a topic, consider when it was published or last revised and whether it will be current enough for your needs.

Finding Books

If you're looking for a specific book or if you're browsing for books on a topic, the best place to start is usually your campus library's website. Most campus libraries offer two types of databases for finding books—one is called a **catalog** and the other is called a **discovery tool**—and you'll learn more about both in Chapter 20. The way these databases are labeled and how you access them varies from library to library. So, this section will provide generic instructions for finding books that should work for most campus libraries.

Suppose you're working on a project for a paleontology class and your instructor recommends that you look at the book *Origin and Early Evolution of the Metazoa.* To check whether your campus library has this book, one option is to search your library's **catalog**. To do that, look on your library's homepage for any links, tabs, drop-down menu options, or search boxes labeled "catalog," "find books," "search books," or something similar. If you see an option like that, use it to search the title of the book. On the library homepage shown in Figure 3.1, the option to search the catalog happens to be in a tab.

Your library may also offer a search box on their homepage that invites you to search for books, articles, media, and more—or has wording similar to that. This is called a **discovery tool** and it offers another way to find books. To look for that book about metazoa, just enter the title into the discovery tool search box, like you see in Figure 3.1. After you've run your search, look for options to filter your results, which are usually listed to the left of your results. Specifically, look for a way to filter your results to *just books*, like in Figure 3.2.

If you don't see options to search either a catalog or a discovery tool on your library's home page, then ask for help. Most libraries make it easy to get help by offering a link to a **chat service** right on the library's homepage. Of course, you can also phone your library, send an email, or visit in person to ask questions.

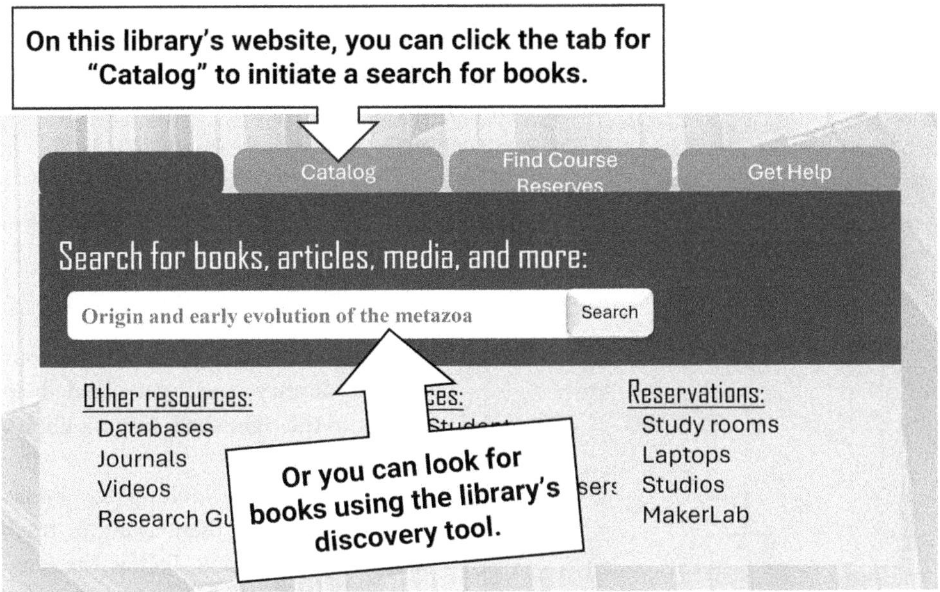

Figure 3.1

Now let's get back to our search. If your library has the book you're looking for, it should appear at or near the top of your results list. If you don't see it, then consult the section on "Dealing with access problems" coming up shortly in this chapter. But let's say you *do* find a record for it (see Figure 3.3).

If this book was available to read online as an e-book, you'd see a link in this record to view the full text online. But this happens to be a **print book**—meaning, it's a physical copy of a book that's printed on paper. You can tell because it has a physical location in the library ("Book Stacks") and it has a call number: QE721.2.E85.75 1992. **Book stacks** just means the shelves in the library and the **call number** is the book's address on the shelf. Figure 3.4 shows how the book looks on the shelf, with arrows pointing to its address (a.k.a. call number).

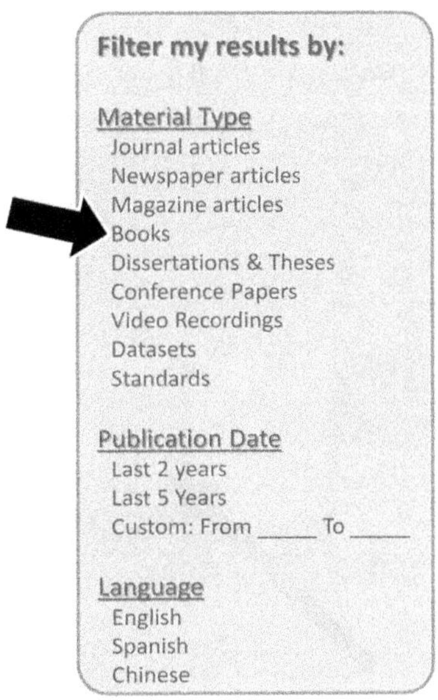

Figure 3.2

Book

Origin and Early Evolution of the Metazoa
Author: Jere H. Lipps
Publication date: 1992
Publisher: Springer

Available in Book Stacks at QE721.2 .E85 075 1992

Figure 3.3

Figure 3.4

If you need to find a print book at your campus library and you need help getting to the right shelf, just ask library staff for assistance.

Suppose you're looking for books on a topic, rather than trying to track down a specific book. Follow the same procedure outlined above, but this time enter keywords for your *topic* into your library's catalog or discovery tool. Let's suppose you're searching for books on the topic of **rammed earth construction**. Figure 3.5 shows the first two results retrieved from that search. The first is an e-book you can view by using the "Click for Online Full-text" link. The second is a print book as evidenced by its physical location (Book Stacks) and call number (TH1421. D48 2020).

You might find that you aren't satisfied with the books your library has to offer, and you want to broaden your search to find more books on this topic. In other words, you want to find out what books exist on this topic, regardless of whether your library owns them or not. To do that you could search an online bookseller like Amazon or you could search Google Books. In Figure 3.6 you see a search of Google Books on the topic of rammed earth construction.

Book

Rammed Earth: Design and Construction Guidelines

Author: Peter Walker, Rowland Keable, Joe Martin, Vasilios Maniatidis
Publication date: 2010
Publisher: IHS BRE Press

Click for Online Full-text

Book

The Art of Earth Architecture: Past Present and Future

Author: Jean Dethier
Publication date: 2020
Publisher: Princeton Architectural Press

Available in Book Stacks at TH1421 .D48 2020

Figure 3.5

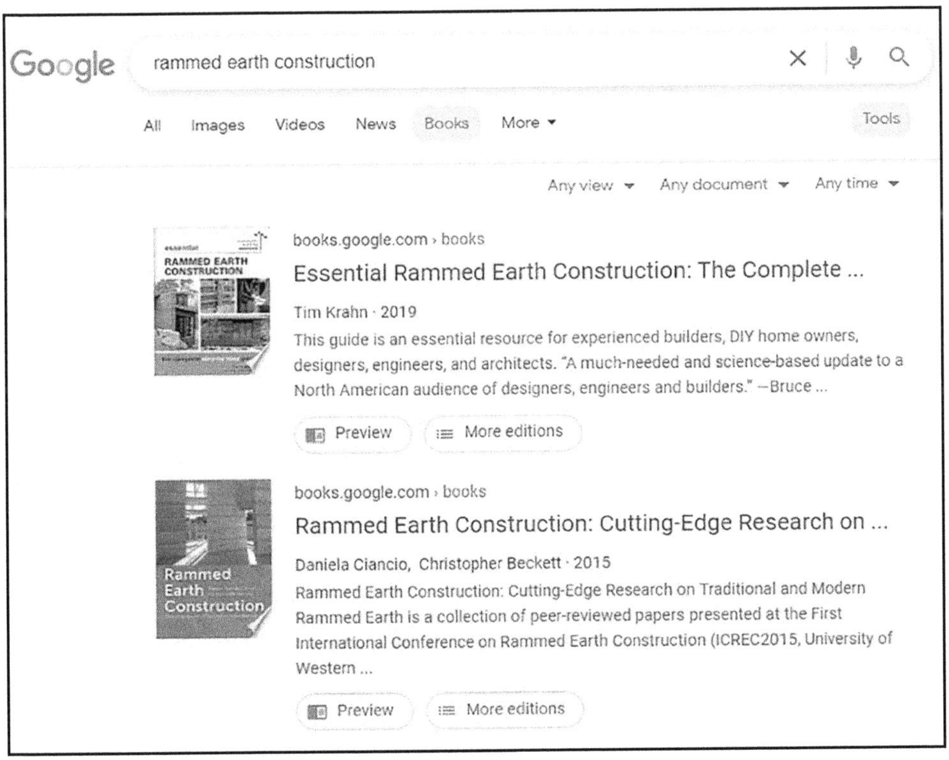

Figure 3.6

You can also look for books in WorldCat, which is a database that combines the catalogs of more than 10,000 libraries around the world. You can easily find WorldCat with a Google search, and you can search it for free. To narrow your search to just books in this database, use the filter menu on the left side of your search results.

Your searches of Amazon, Google Books, and/or WorldCat may turn up books that look promising to consult, but without any way to access those books for free. In that case, note down the titles and search your library's catalog or discovery tool for them. If you can't find them, try searching Google to see if you can find a free version to access somewhere on the internet. But if that doesn't work either, then consult the section below on "Dealing with access problems."

Accessing Free, Online Books

As you just learned, you can try searching Google (or Google Scholar) for a specific book to see if you can find an online copy that's free to read. To search, try entering the book's title along with the last name of the author.

The internet also provides collections of **open-access books**, which are published online—often by the author—and made freely available for anyone to read. Below are a couple of databases for finding open-access books; these databases can easily be found with a Google search:

- Directory of Open Access Books (DOAB)
- JSTOR Open Access E-Books

To find more resources directing you to open-access books, just do either or both of these Google searches:

- **open-access books**
- **open-access books libguides**

Some open-access books are *textbooks*, and there are databases devoted to finding them such as:

- Open Textbook Library
- OER Commons

Open-access textbooks are also known as **open educational resources**. To find more resources directing you to open educational resources, try any or all of these Google searches:

- **open educational resources**
- **OER textbooks**
- **open textbooks**

In addition, some databases help you find any kind of open-access book (including open-access textbooks) as well as other open-access content. Examples include:

- Openly Available Sources Integrated Search (OASIS)
- Mason OER Metafinder (MOM)

Just so you know, some books are in the **public domain**—meaning they were published a long time ago as print books, but they're too old to still be protected by copyright law, so they've been digitized and made available for free on the internet. Databases for finding them include:

- Biodiversity Heritage Library
- HathiTrust
- Google Books

To find more resources directing you to free, online, public domain books, try either or both of these Google searches:

- **Public domain books**
- **Public domain books libguides**

Dealing with Access Problems

So, you want to consult a particular book, but your campus library doesn't seem to have it, and when you search for it on the internet, you can't find a free version to read. In that case you can take advantage of a service your library offers for getting books another way. Depending on the library, this service might be called either **interlibrary loan** or **document delivery.** Regardless of the name used, when you submit a request for a book using this service, your library will find another library that has the book and ask that library to lend their copy. When the book arrives at your library, you'll get a notice to come pick it up. Typically, you get to keep the book for about three weeks before your library has to send it back to the library that lent it, but often you can request to renew it so you can keep it longer.

You should see a link on your campus library's homepage leading to their interlibrary loan/document delivery service. But, if you need help finding and/or using this service, consult library staff.

If you're thinking that your library's document delivery/interlibrary loan service would be a great way to request your course textbooks for free—sadly, it isn't. Most campus libraries don't collect course textbooks and obviously they can't lend what they don't have. So, if you put in a document delivery/interlibrary loan request for one of your textbooks, it'll likely come back unfilled.

Distinguishing Books from Other Sources

Suppose you're searching a library database and you aren't sure if it's presenting you with results that are books. In that case, you can look for filters to refine your results to *just* books (you saw an example of how to do this earlier in the chapter), or you can check to see if the database identifies books as a source type in your results. For example, Figure 3.7 shows some database results from a search on how biocrusts deter dust storms. Notice how each result is labeled by source type at the top of the record, and there are two books in these results. One is available online, the other is a print book available in the library.

Magazine article

Biocrusts Keep the Dust Settled
Author(s): Ogasa, Nikk
Science News Volume 201, Issue 11, Pages 12-13
June 2022

Click for Online Full-text

Scholarly journal article

Soils' Coating Helps Keep Dust Storms at Bay
Author(s): Pennisi, Elizabeth
Science Volume 376, Issue 6595, Pages 786-787
May 2022

Click for Online Full-text

Book

Biological Soil Crusts: An Organizing Principle of Drylands
Author(s): Weber, Bettina, et al.
2016
Springer Publishing

Click for Online Full-text

News article

A Microscopic Issue of Unknown Consequences
Author(s): Fountain, Henry.
New York Times 2014

Click for Online Full-text

Book

Biological Soil Crusts: Structure, Function, and Management
Author(s): Belnap, Jayne & Lange, Otto
2001

Available in Book Stacks at QH84.8.B555 2001

Figure 3.7

When books are clearly labeled in your results (like you see in Figure 3.7) there's no ambiguity about the type of source you're looking at. But depending on what database you're searching, your results may not be labeled by source type. In that case, examine the source for the following clues to determine if it is a book:

- Check the title of the source for "bookiness." Typically book titles indicate you'll get comprehensive coverage of a topic.
- Look to see if the source is divided into **chapters**.
- Look for a **front cover** and **title page**; both should be present regardless of whether the book is in print or online. (A title page is a page near the front of the book that shows the book's title, author, and publisher.)

Of course, whenever you are in doubt about the type of source you're looking at, you can always ask library staff for help!

To sum up, what should science and engineering students know about books?

1. Books explain a topic at length (excluding reference books). So, when you want to thoroughly educate yourself on a topic, a book is what you'd reach for.
2. Knowing there are different types of books (such as popular books, scholarly books, textbooks, and reference books) allows you to zero in on the best type of book for your information need.
3. Books tend to be less current than other types of sources.
4. Books are available from libraries, as well as through library services such as document delivery/interlibrary loan; some books are also freely available on the internet.

What's Next?

In the next chapter we'll take a close look at **newspapers** and other news sources, and you'll learn how science and engineering topics can be distorted or misrepresented in news articles.

CHAPTER 4
NEWSPAPERS AND OTHER NEWS SOURCES

Are you wondering why *news* is important to scientists and engineers? This chapter will explain! Not only will you get an overview of different news sources, you'll learn how science and engineering news is skewed towards certain topics and that it's often misrepresented. This chapter will also explain where to find news articles and what to do if you can't access a news article you need.

An Overview of Newspapers

When new and noteworthy events occur, such as political elections, epidemics, natural disasters, or breakthroughs in science or engineering, we call it **news**. Newsworthy events are written up by journalists and published as **newspaper articles** in newspapers. The intent of news articles is to keep the general public informed about important current events, and the writing style is simple and clear so readers can easily understand the content.

The first newspapers started in Europe in the 1600s. They grew in popularity and abundance worldwide until the emergence of the internet. Since then, they've declined in numbers as more people opt to get news from other types of sources available on the internet. According to the Newspaper Association of America, there are around 1,000 daily newspapers currently published in the United States today. There are also many newspapers that used to exist but have ceased publication.

Most newspapers are published daily and so they are sometimes referred to as **dailies**. Each daily installment is called an **issue** or **edition**. In the United States there are **national dailies** like *U.S. News and World Report*. There are also **metropolitan dailies** like *The Washington Post* and *The New York Times*. These metropolitan dailies are based in a city and provide national and international news as well as news local to that city. Furthermore, a few dailies are devoted to **business news** such as *The Financial Times* and *The Wall Street Journal*.

Other News Sources Include Radio, TV, Magazines, and News Websites

It used to be that newspapers were the main source of news. But as time passed, news was communicated in more formats including magazines (such as *Time* and *Newsweek*), radio stations (such as NPR, CBS News Radio, etc.), and TV (such as CNN, NBC, FOX,

BBC, Al Jazeera, etc.). When the internet came along, all these pre-existing print and media news sources developed websites so they could simultaneously share (or sell) their news through the internet.

As more time passed, lots of news websites were "born" on the internet, meaning they didn't start out as print newspapers, magazines, or TV/radio stations. We consider these news sources to be **born-digital**, because they didn't exist in any other form before the emergence of the internet. Examples of born-digital news sources include Slate, HuffPost, ProPublica, Politico, The Hill, etc. There are also born-digital news websites that specialize in delivering **science and engineering news**. A good example is ScienceDaily.com, but a Google search on a particular research area along with the word "news" will reveal others.

So, there are lots of news sources and some have a long history (such as newspapers), whereas others have a short history (such as "born-digital" news websites).

The Spectrum of Content Found in News Sources

While news sources report on events happening in the world, most contain other types of content as well. For example, most news sources present readers with news analysis as well as opinion pieces. **News analysis** provides an interpretation of news events from an expert who can weigh in on the meaning, implication, or significance of those events. **Opinion pieces** offer someone's opinion on a topic and examples include editorials, commentaries, op-eds, columns by regular or guest columnists, letters to the editor, book reviews, movie reviews, etc. Here's an example of an opinion piece about smart phones that appeared in *The New York Times*:

- Beres, D. (2021). Your smartphone should be built to last. *New York Times*. https://www.nytimes.com/2021/04/25/opinion/iphone-apple-electronics-technology.html

Keep in mind that opinions are not really news, even though they appear in news sources. Rather, they represent someone's perspective on a topic. For example, in that *New York Times* opinion piece about smartphones, the author is deploring how constantly upgrading smartphones results in a lot of waste, plus it's expensive for consumers:

> Consumers are urged to upgrade their devices annually. Well north of 1 billion smartphones were shipped in 2020—and it was a sluggish year because of the Covid-19 pandemic. Manufacturers must do better. Their devices must be repairable by all and kept compatible with software updates for as long as possible, not artificially obsoleted. Consumers should support right-to-repair legislation.

News sources may also include content intended to entertain readers, such as crossword puzzles, comics, recipes, celebrity gossip, etc. And some news sources barely deliver any real news at all, but instead churn out sensationalized articles that we refer to as **click-bait**, like this article:

- How using Facebook could increase your risk of cancer. (2009). *Daily Mail.* https://www.dailymail.co.uk/health/article-1149207/How-using-Facebook-raise-risk-cancer.html

Biased News Reporting

You may have heard that *all news* is biased toward a liberal or conservative perspective, but that's not strictly true. **Bias** is prejudice toward someone or something in a way that's unfair or inaccurate. In journalism, bias can result in news being unreliable, unbalanced, misleading, or even fabricated. However, a lot of science and engineering news is *not* subject to bias—for example, news reports on the discovery of a new animal species or about the Hubble Space Telescope are unlikely to be biased. That's because those topics are not political, nor have they become politicized. On the other hand, there are some science and engineering topics that *have* become politically polarized, such as climate change and stem cell research. So, you may encounter biased coverage on those topics.

But as a lot of news *is* political (or has become politicized), how do news sources deal with bias? Some news sources embrace and promote bias as this makes them appealing to a niche audience who want to hear news that justifies the beliefs they hold. Other news sources strive to be objective and neutral, which usually means they try to provide their audience with a balance of liberal and conservative perspectives.

Of course, there's no news source that achieves a perfect balance all the time. Plus, it's not clear that news sources should *always* strive to be neutral and balanced. For example, is it appropriate to give equal time to two opposing viewpoints when one of those viewpoints is not supported by any evidence?

Where Should You, as a Future Scientist or Engineer, Get Your News?

The fact is that some news sources *are* better than others, and if you're interested in facts, as all scientists and engineers should be, you should seek news from high-quality sources. To explore what's out there, you might try searching Google for **media bias charts.** That search will lead you to resources that rank news sources across spectra such as reliability and bias. If you decide to consult charts like this, keep in mind that many of the news sources that rank highly for accuracy, reliability, and objectivity have still made serious errors in reporting over the years. No news source is perfect or ever will be!

You can also try visiting various fact-checking websites to verify the accuracy of the news you're reading. To find these sites, just do a Google search on **fact-checking websites**, and you'll find resources like FactCheck.org, Politifact.com, Snopes.com, etc.

Science and Engineering in the News

It's usually in news sources that the general public learns about advances in science and engineering, so this is where scientists and engineers tend to get the most exposure. This has a few implications, none of which are terribly positive. First, the news that journalists report on is not an even distribution of discoveries across science and engineering disciplines. Instead, what makes the news are advances likely to be of interest to the general public, such as research on diet, health, personal technologies, etc.

Second, journalists don't always fully understand the research they're reporting on, so they get facts wrong. This is not surprising because advances in science and engineering can be very complicated and difficult to comprehend unless you're an expert.

Third, journalists sometimes misunderstand how reliable and conclusive a research study's findings are. Many research studies are designed to be exploratory in nature—plus, researchers usually don't reach a consensus that they've found the right answer to a question until multiple research studies have been conducted that replicate and confirm results. When journalists aren't aware of this, they may report tentative results as facts. However, sometimes journalists do this on purpose. That is, they'll report on exciting new science and engineering research results simply to attract readers—but without considering how preliminary or exploratory those research results are. Here are some interesting research articles that have studied this problem:

Dumas-Mallet, E., et al. (2018). Scientific uncertainty in the press: how newspapers describe initial biomedical findings. *Science Communication*, *40*(1): 124–141.

Amberg, A. & Saunders, D. N. (2020). Cancer in the news: bias and quality in media reporting of cancer research. *PLoS One*, *15*(11): e0242133.

Oxman, M., et al. (2021). Quality of information in news media reports about the effects of health interventions: systematic review and meta-analyses. *F1000Research*. DOI: 10.12688/f1000research.52894.2.

Dumas-Mallet E., et al. (2017). Poor replication validity of biomedical association studies reported by newspapers. *PLoS One, 12*(2): e0172650.

The conclusion section of that last study provides a nice summary of the problem:

> Our study shows that many biomedical findings reported by newspapers are disconfirmed by subsequent studies. This is partly due to the fact that newspapers preferentially cover "positive" initial studies rather than subsequent observations, in particular those reporting null findings. Our study also suggests that most journalists from the general press do not know or prefer not to deal with the high degree of uncertainty inherent in early biomedical studies.

The bottom line is if you want to promote your research, and you're working with a journalist, reporter, or blogger, you need to be very careful the person you're working with understands your research and writes it up accurately. If you are working with your employer's public relations department to publicize your research in a press release, *be just as careful.* Press releases can also distort or exaggerate research findings and you'll see an example of this in Chapter 16.

Finding Newspaper Articles

If you need a specific newspaper article or if you're browsing for newspaper articles on a topic, the best place to start is usually your campus library's website. Look for a search box on your library's homepage that invites you to search articles, books, media, and more—or has wording similar to that. This is called a **discovery tool** and you'll learn more about it in Chapter 20. You saw how to search a discovery tool for a particular book in Figure 3.1, but you can also look for titles of newspaper articles using this search box, or you can enter a topic and then once you've run your search look for filters (which are usually to the left of your results) to limit your results to newspaper articles only. You learned how to filter to books in Figure 3.2, so just use the same strategy but instead filter to newspapers.

If your library doesn't provide a discovery tool search box on their homepage, or if the search you conducted didn't find what you needed, then your best bet is to ask for help. You can look for a link to a **chat service** on your library's homepage and initiate a chat to ask how to find newspaper articles. Or you can phone or email your library to ask, or visit in person.

If you feel like being more self-sufficient before asking for help, then you can also consult Chapter 20 for tips on navigating your library's website to find **content-specific databases** that specialize in providing newspaper articles, such as Access World News, ProQuest Newspapers, and Gale OneFile: News.

Or you can try searching Google to find a specific news article. Suppose someone told you about an article in *The Wall Street Journal* written by David Gompert about the use of quantum communication for cybersecurity. To find it, try searching Google with a mix of keywords that represent the author, what the article is about, and where it was published, like this: **david gompert cybersecurity wall street journal**

You could also go directly to the website for *The Wall Street Journal* and search for the article there. However, even if you find the article, when you try to read it you might encounter a notice asking you to pay first. This is called a **paywall**. If that happens, then check to see if you can get the article from your library instead, using the techniques explained at the beginning of this section. If you can't access the article through your library, then ask for help or consult the section below on "Dealing with access problems."

Browsing Newspaper Articles

You can also browse through the online version of a newspaper to find articles, but you may find you can only read a certain number of articles for free before you're asked to pay. The online version of a newspaper is usually presented on a website run by the newspaper's publisher. To find it, do a Google search on the title of the newspaper.

Note that the online version of the newspaper will have the same content as the print version. However, it is important to know that articles in the online version may have slightly different titles than those that appear in the print version. For example, a March 11, 2017 *New York Times* article appears in print under the headline "The State-by-State Assault on Electric Cars," while the online article is published as "Behind the Quiet State-by-State Fight Over Electric Cars." In the online edition, there is a footnote at the end of the article that states:

> A version of this article appears in print on March 12, 2017, Section BU, Page 1 of the New York edition with the headline: The State-by-State Assault on Electric Cars.

Whereas the online version of a newspaper will have the same content as the print version, it may also provide additional content not found in the print version, such as breaking news, supplementary articles, updates to previously published articles, etc.

If you feel like browsing through a print issue of a newspaper, then be aware that many libraries subscribe to their local city or town paper, as well as a few other major papers like *The New York Times* and *The Wall Street Journal*. Simply go to your library and ask where current, print newspapers are located.

Dealing with Access Problems

Sometimes there's a newspaper article you want to read, but your library can't provide access to it, nor can you find a free copy to view on the internet. To get the article, use the same service you learned about in the previous chapter, which will either be called **interlibrary loan** or **document delivery**. You should be able to find a link to this service on your campus library's homepage. Once there, you can enter the details for the newspaper article you need and submit your request. Your library will find another library that can supply the article so you can read it. If you need help finding and using your library's interlibrary loan/document delivery service, just ask library staff for assistance.

Distinguishing Newspaper Articles from Other Sources

Newspapers are easily recognizable in their physical form (which we refer to as the **print format**). They are usually large, folded, and have a distinctive layout. Plus, they're printed on thin paper called **newsprint**. But distinguishing newspaper articles from other sources in an *online* environment can be harder, so let's go over some strategies for doing that.

Suppose you're searching a library database and you aren't sure if it's presenting you with results that are newspaper articles. You can look for filters to refine your results to *just* newspaper content, or you can check to see if the database identifies newspaper content in your results. For example, in the last chapter you saw database results for a search on biocrusts (see Figure 3.7). One of those results is clearly labeled as a news article. When newspaper articles are clearly labeled like this, there's no ambiguity about the type of source you're looking at. But depending on where you're searching, your results may not be labeled by source type. For example, suppose you're researching **coral reef conservation** and you conduct a Google search to find sources. Your Google results won't be labeled by source type, but some of them might be newspaper articles, such as the third result shown in Figure 4.1.

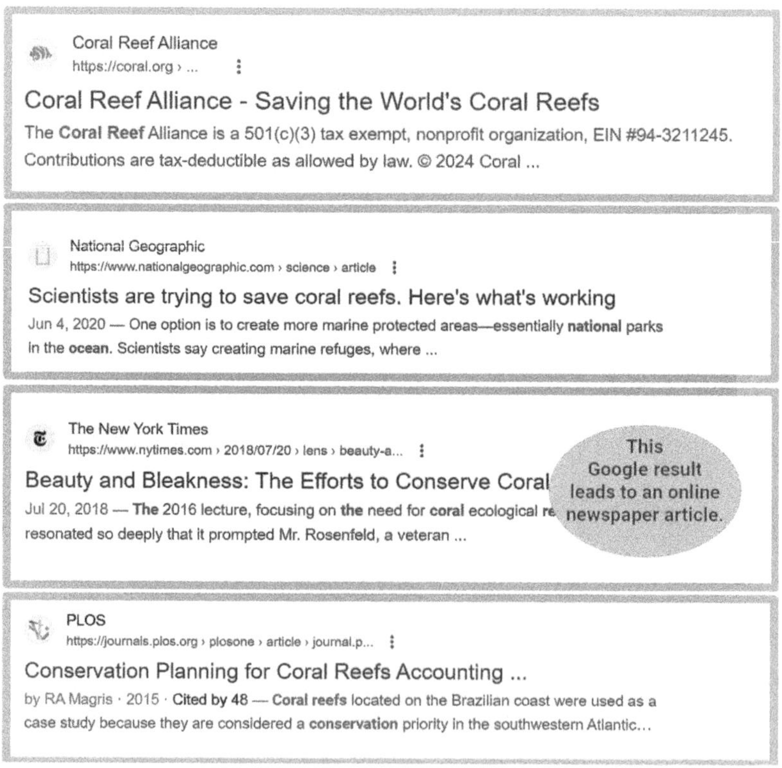

Figure 4.1

How can you tell the "Beauty and Bleakness" result leads to a newspaper article? Recall from Chapter 2 that experienced researchers can easily pick up on clues in their Google results that help them identify source types. In this case, the result indicates it's coming from *The New York Times*, a well-known newspaper.

If you have trouble picking up on source-type clues within your Google results, you may find more obvious clues once you click a result and view the source. Here are some characteristics that might help you distinguish online newspaper articles from other sources (also see Figure 4.2).

- Look at the top of the page, above the title of the article, and see if you notice a heading that sounds "newspapery." For example, metropolitan dailies usually have their city name in the title followed by words like **tribune**, **post**, **dispatch**, **times**, **herald**, **sentinel**, **gazette**, **standard**, **chronicle,** or simply **news**.
- Look around for navigation that leads to typical newspaper sections such as: World news, U.S. News, Politics, Business, Opinion, Obituaries, Sports, etc.
- Look for the presence of a search feature.
- Look for links that invite you to subscribe and/or sign in.
- Try looking up the name of the source (located above the article title) in Wikipedia, to see if that helps you determine what type of source it is.

Of course, whenever you are in doubt about what type of source you're looking at, you can always ask library staff for help!

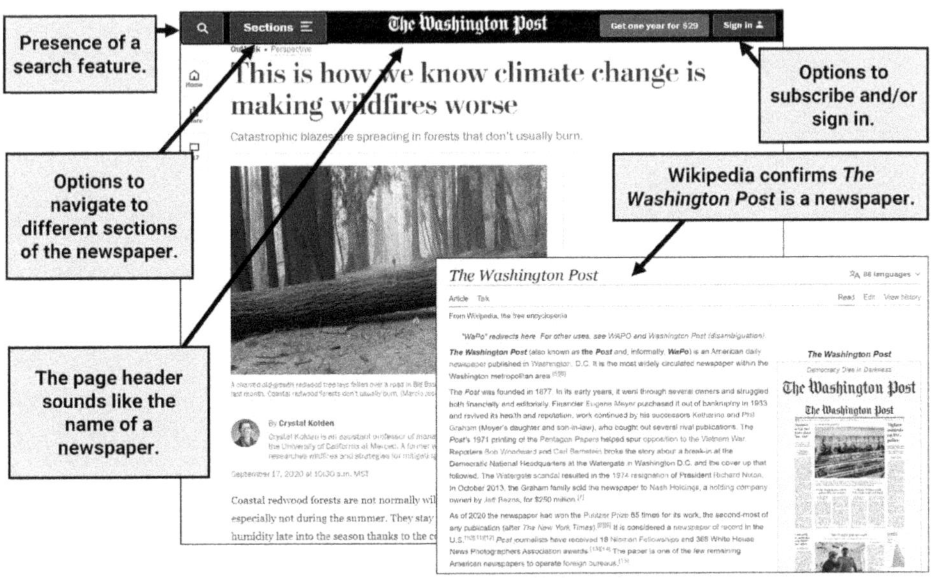

Figure 4.2

To sum up, what should science and engineering students know about news sources?

1. There are different types of newspapers and news is also presented in magazines, websites, radio, TV, etc.
2. Many news sources provide content other than just news articles, such as news analysis, opinions, entertainment, etc.
3. Some news sources are consistently—and even proactively—inaccurate and/or biased, so you'll need to choose your news sources carefully.
4. Science and engineering news tends to be skewed towards topics that will be of interest to the general public.
5. Journalists are prone to making errors when reporting on science and engineering news; for example, they may report tentative results as facts.
6. Newspaper articles can be found on the internet, from library databases, as well as through library services such as document delivery/interlibrary loan. Libraries provide staff to help you locate newspaper articles when you need assistance.

What's Next?

In the next chapter you'll learn all about **magazines**, including **trade magazines** which are regularly used by some scientists and engineers.

CHAPTER 5
MAGAZINES

Many scientists and engineers consult trade magazines, so if you're not familiar with them, this chapter will fill you in. In fact, you'll learn about a variety of magazine types in this chapter, as well as where to search and find magazine articles, and what to do if you can't access a particular article you need. Finally, you'll learn how to distinguish online magazine articles from other online sources.

An Overview of Magazines

Magazines can be divided into two categories: **popular magazines** and **trade magazines**. Both consist of **articles** on various topics, and both are printed on paper and delivered to paying subscribers in installments called **issues**—usually weekly, monthly, or quarterly. Like newspapers, the first magazines started in Europe in the 1600s. According to the Statista website, in the United States there are over 7,000 different magazines currently published. And, of course, there are lots of magazines that used to exist but have ceased publication.

Popular magazines are written for a general audience and are easy for anyone to read and understand. Their purpose is to entertain, educate, inform, persuade, or any combination of these. Many popular magazines focus on a particular interest or hobby such as interior design, autos, electronics, sports, travel, crafts, cooking, gardening, fashion, etc. Examples include:

- *Country Living*
- *Car and Driver*
- *Sports Illustrated*
- *Wired*
- *Travel and Leisure*

Some popular magazines are devoted to business and finance, for example:

- *The Economist*
- *Forbes*
- *Fortune*
- *Entrepreneur*
- *Bloomberg Business Week*

Other popular magazines provide news articles, such as:

- *Time*
- *Newsweek*
- *The Week*
- *The National Review*
- *Foreign Affairs*

And some popular magazines cover science and engineering, such as:

- *Scientific American*
- *Air and Space*
- *Discover*
- *Sky and Telescope*
- *Popular Science*

Meanwhile, **trade magazines** contain articles of interest to professionals working in a particular trade or industry. A trade magazine might include news and product information, latest best practices, market trends, regulatory or policy issues, or any other information that would be particularly helpful for readers wanting to keep up with that profession. Here are some examples:

- *IEEE Communications*
- *InTech*
- *Chemical Week*

IEEE Communications helps professionals involved with electronic media keep up with new technologies and market trends. *InTech* helps engineers in automation technologies keep up to date, and *Chemical Week* provides chemists with industry news.

If you have plans to work in a certain industry or profession, you might want to check whether there are trade magazines that would be useful to subscribe to. How would you check? Just run some Google searches on the name of your profession along with the words **trade magazines**. Note that trade magazines are sometimes called **trade journals**, **professional magazines**, or **professional journals**, so you might want to substitute in some of those name variations when you run searches in Google.

Because some trade magazines are published by professional societies, another way to find them is to go to the website for a professional society and look for any trade magazines they publish. For example, if you went to the website for the American Institute of Chemical Engineers (AIChE), you could click their link to "Publications" in the top menu bar and find information about their trade magazine called *Chemical Engineering Progress (CEP) Magazine* (https://www.aiche.org/publications/cep).

Finding Magazine Articles

If you need a specific magazine article or if you're browsing for magazine articles on a topic, the best place to start is usually your campus library's website. Look for a search box on your library's homepage that invites you to search articles, books, media, and more—or has wording similar to that. This is called a **discovery tool** and you learned about it in the previous two chapters, and you'll learn even more about it in Chapter 20. To find a specific magazine article, try searching its title, or you can enter keywords to look for magazine articles on a topic. Once you've run your search, look for filters (which are usually to the left of your results) to limit your results to magazine articles only.

If you don't see a search box like this on your library's homepage, or if the search you conducted didn't find what you needed, then ask library staff for help. Look for a link to a **chat service** on your library's homepage and initiate a chat to ask how to find magazine articles. Or you can phone or email your library to ask, or visit in person.

However, if you feel like exploring a bit more on your own, then consult Chapter 20 for tips on navigating your library's website to find **content-specific databases** that specialize in providing magazine articles, such as Gale OneFile: Popular Magazines. Magazine articles can also be found in **article databases** like Academic Search, ProQuest Central, and OmniFile Full Text.

Or, you can try searching Google to see if you can find a specific magazine article. Searching the article's title is usually the most effective strategy. You can also go to a particular magazine's website and search for articles there. To find a magazine's website, just search Google for the magazine's title. It may also help to add the word "magazine" to your search. If you try to access an article but get a notice asking you to pay (called a **paywall**) then go to your library's website and look for the article there, using the techniques explained above. If you still can't find it, ask for help or consult the section below.

Dealing with Access Problems

If there's a magazine article you want to read, but your library doesn't seem to have it, nor can you find it for free on the internet, then request it through your library's **document delivery/interlibrary loan** service. The previous two chapters explained this service, but if you are confused about how to use it, then ask library staff for assistance.

Distinguishing Magazine Articles from Other Sources

Magazines in their print form are fairly distinctive—they're usually printed on glossy paper and have lots of colorful photos, illustrations, and advertisements. But distinguishing magazine articles from other sources in an *online* environment can be harder, so let's look at tips for doing that.

Suppose you're searching a library database and you aren't sure if it's presenting you with results that are magazine articles. One thing you can do is look for filters to refine your results to *just* magazine content. Or, you can check to see if the database identifies magazine articles in your results. Look again at those search results on biocrusts shown in Figure 3.7. You'll see the very first one is a magazine article.

But not all databases label results by source type; for example, Google doesn't. Remember that Google search **on coral reef conservation** from the previous chapter? If you recall, it turned up a newspaper article, but it also turned up magazine article; it's the second result in the image in Figure 5.1.

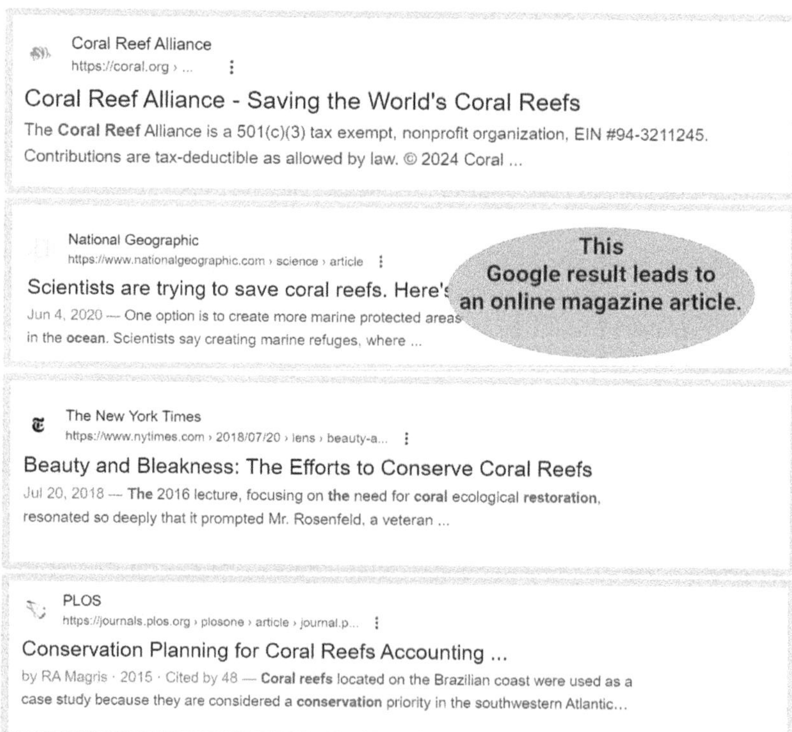

Figure 5.1

Notice that result is coming from *National Geographic*, which many people recognize as the name of a popular magazine. But if you didn't know that, don't worry! As time goes on, you'll get more familiar with different magazine titles and be able to recognize those titles in your Google results.

If you have trouble picking up on source-type clues within your Google results, you may find more obvious clues once you click a Google result and view the actual source. Here are some characteristics (Figure 5.2) to look for to distinguish magazine articles from other sources:

- Look for the presence of a search feature.
- Look for links that invite you to subscribe and/or sign in.
- Try looking up the name of the source (located above the article's title) in Wikipedia, to see if that helps you learn what type of source it is.

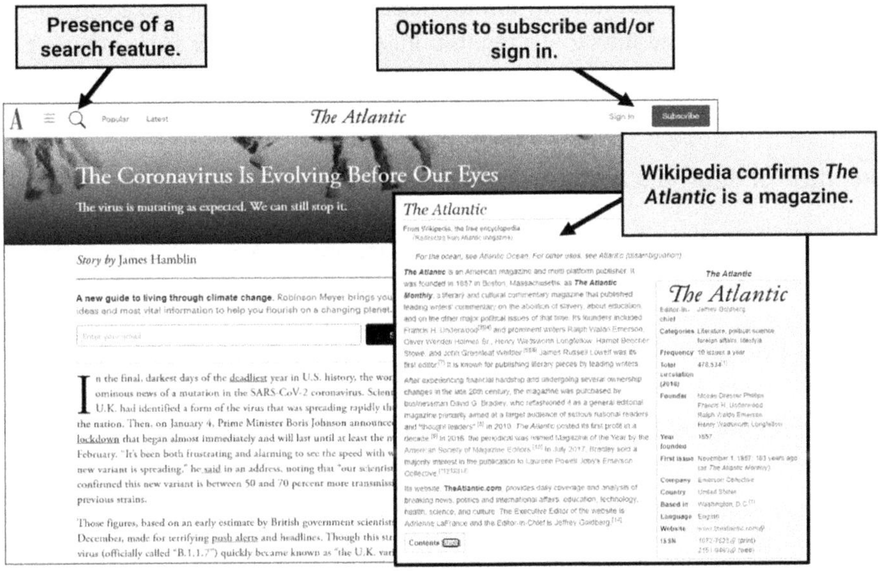

Figure 5.2

To sum up, what should science and engineering students know about magazines?

1. There are two main types of magazines: popular magazines and trade magazines.
2. Many professionals, but especially engineers, use trade magazines to keep up to date in their profession.
3. Magazine articles can be found on the internet, from library databases, as well as through library services such as document delivery/interlibrary loan. Libraries provide staff to help you locate magazine articles when you need assistance.

What's Next?

In the next chapter you'll learn what **journals** are and why they are especially important sources for scientists and engineers.

CHAPTER 6
JOURNALS AND JOURNAL ARTICLES

Journals, and the articles they contain, are especially important sources for scientists and engineers. This chapter will help you understand what journals are and what you can expect from the articles found within journals. This chapter will also cover where to find journal articles and how to distinguish online journal articles from other types of sources.

An Overview of Journals and Journal Articles

You've seen that books, newspapers, and magazines are sources that serve different purposes. Journals are another type of source, and their main purpose is to provide articles describing new discoveries, applications, or inventions that resulted from research. Journals are a particularly important source for scientists and engineers, so we'll explore them in detail, starting with some history.

The first journals were launched in the 1660s and they were created to meet certain needs that researchers had at that time, and which they still have today:

- Research results need to be shared so that others can learn from them and build upon them.
- Research results need to be shared as quickly as possible, so that researchers don't duplicate studies or embark on studies based on obsolete information.

Journals were created to resolve these needs. Prior to the creation of journals, books were the most common type of information source, but books are not ideal for quickly and continually sharing research results. Research results can usually be written up in just a few pages, but books tend to be comprehensive and thus lengthy, so they take a while to write. If researchers had to wait to publish their research findings only after they had enough material to fill a book, this would slow down communication and impede progress.

It turns out that a great way to share research results both quickly and continually is to create a publication that comes out in installments—called **issues**—with each issue containing several individual, short **articles**. We call this type of publication a **periodical** or **serial** because it comes out periodically or serially—either daily, monthly, quarterly, or annually. Journals are a type of periodical publication, as are newspapers and magazines, but journals have their own distinct purpose and that is to continually present the latest research findings so scientists and engineers can refer to each other's work and keep up to date.

The first English-language journal began publication in 1665 and had a really long title: *Philosophical Transactions, Giving Some Account of the Present Undertakings, Studies, and Labors of the Ingenious, In Many Considerable Parts of the World.* Thankfully, journal titles are much shorter now. Many journals have titles that start out "Journal of," as in *Journal of Fish Biology*. Some have one-word titles like *Blood* or *Lung*. Others have acronym or initialism titles like *PLoS One*, *PNAS*, or *JAMA*.

While the earliest journals contained articles about research in science (or **natural philosophy**, as science was called at that time), over time new journals were created to disseminate research results in all sorts of disciplines other than just the sciences. Today there are journals dedicated to research in engineering, sociology, psychology, philosophy, education, history, etc. Various sources list the total number of journals currently in publication at around 30,000.

Most journals are expensive, so individual researchers may subscribe to just one or two low-cost journals, or they may get access to a few journals as a membership benefit when they join a professional society. However, many researchers don't subscribe to *any* journals. Rather, campus libraries are the main subscribers because subscriptions are so expensive. To learn more about why journals are so expensive, consult Appendices B and C.

Content You'll Find in a Journal

Journals primarily consist of articles and there are two main types. **Research articles** describe research projects and what was learned from those projects, while **review articles** provide an overview of what we know about a specific area of research. (You'll learn more about these two types of journal articles in the next couple of chapters.)

To better understand what journals are and how they differ from other sources, it might help to compare them with newspapers:

<table>
<tr>
<td>

• A **journal** is a publication containing **articles** about **research projects** that have taken place.
• Journal articles are written by **researchers**.
• Most journals are published in **monthly issues**.

</td>
<td>

• A **newspaper** is a publication containing **articles** about **newsworthy events** that have taken place.
• Newspaper articles are written by **journalists**.
• Most newspapers are published in **daily issues.**

</td>
</tr>
</table>

The writing style of journal articles is formal, neutral, and factual, as well as chock-full of technical terminology. So, unlike newspaper articles, journal articles are not written for the general public to easily digest; rather, the audience is other researchers.

Many journals provide other types of content in addition to research and review articles, such as news, editorials, letters to the editor, book reviews, commentaries, protocols, etc. Table 6.1 compares content offered in different journals. Some journals, like *Geology*, just provide articles in each issue, while other journals provide articles along with a mix of other content.

Table 6.1

Comparison of content offered in different journals			
New England Journal of Medicine	*Journal of Environmental Engineering*	*Geology*	*Nature Ecology and Evolution*
Perspective Original articles Review articles Editorials Correspondence	State-of-the-art reviews Technical articles Technical notes	Articles	Editorial Correspondence Comment & opinion News & views Review articles Research articles

Journal Issues and Volumes

Journal issues typically have anywhere from three to twenty articles per issue, and a whole year of issues is called a **volume**. But let's slow down and examine issues and volumes in more detail. Suppose that in the year 2021, you started a journal called *Journal of Brand-New Research* and you published one issue a month.

In January you'd put out issue 1, in February issue 2, and so on—with your last issue being number 12 for December. That entire year of issues would then make up volume 1 of the journal. The next year (2022) you'd put out another twelve monthly issues, numbered from 1 to 12, and those issues would comprise volume 2. In 2023 you'd put out another 12 issues, and those would make up volume 3, and so on. Below you see the cover of volume 3, issue 12 of *Journal of Brand-New Research* (Figure 6.1).

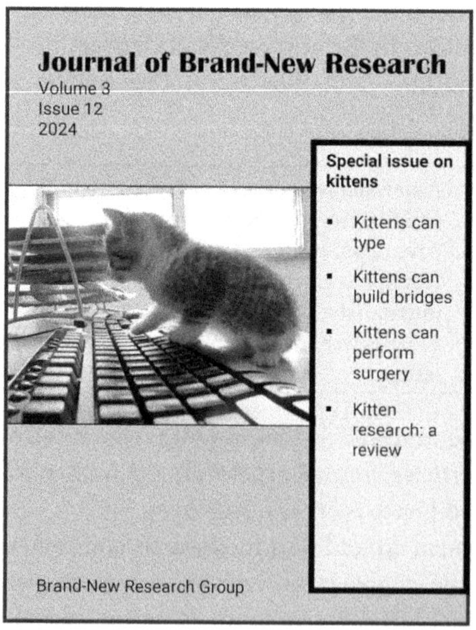

Figure 6.1

Note that some journals come out bimonthly or quarterly instead of monthly. Regardless of the frequency with which issues are released, at the end of each year the issues released comprise a **volume**. So, if you're ever wondering how long a journal has been around, note that if volume 3 was published in 2023, then you can calculate back to determine that volume 1 was published in 2020.

Volume and issue numbers are important to know about because they show up in **citations**. You'll learn more about citing and citations in Chapters 29–31, but for now just be aware the volume and issue number appear toward the end of the citation, just before the page numbers. Here's an example citation (Figure 6.2) from our make-believe journal.

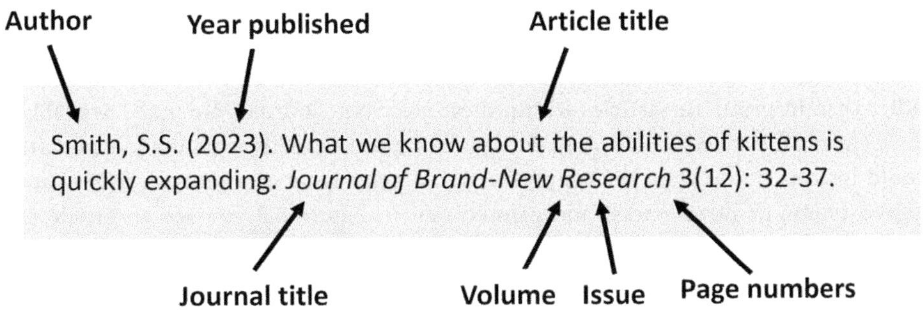

Figure 6.2

In the past, campus libraries would receive print journal issues throughout the year in the mail. When each issue arrived, it would be made available to library users to browse through. Then, at the end of the year, the library would collect all those print issues and mail them off to a company that would bind them together into a single volume. The name of the journal and its volume number would be printed on the spine of each volume. These bound volumes would be sent back to the library and shelved with volumes from previous years. So, that's what you're seeing if you wander around a campus library and notice shelves filled with lots of consecutively numbered volumes.

However, nowadays campus libraries subscribe to most journals in their *electronic* form (which are sometimes called **e-journals**), because it's a lot more convenient and it saves shelf space. So, you will rarely—if ever—be accessing journals in their print form. In fact, as there's very little demand for print journals, some publishers have shifted to publishing their journals online only and have abandoned the print format altogether. Also, because e-journals have largely replaced print journals, it's common for people to drop the "e" and simply refer to "e-journals" as "journals."

Journal Article PDFs

You just learned that nowadays most journals are accessed in their electronic form, rather than in print. The articles within these electronic journals are made available as an electronic file called a PDF. A **PDF** digitally captures exactly what the journal article would look as if you were thumbing through a print issue of a journal. PDFs can be viewed online or downloaded and printed out. In Figure 6.3 you see an article as it appears within a print journal issue, along with the same article downloaded as a PDF, then printed out and stapled together. Notice that the layout and text are identical.

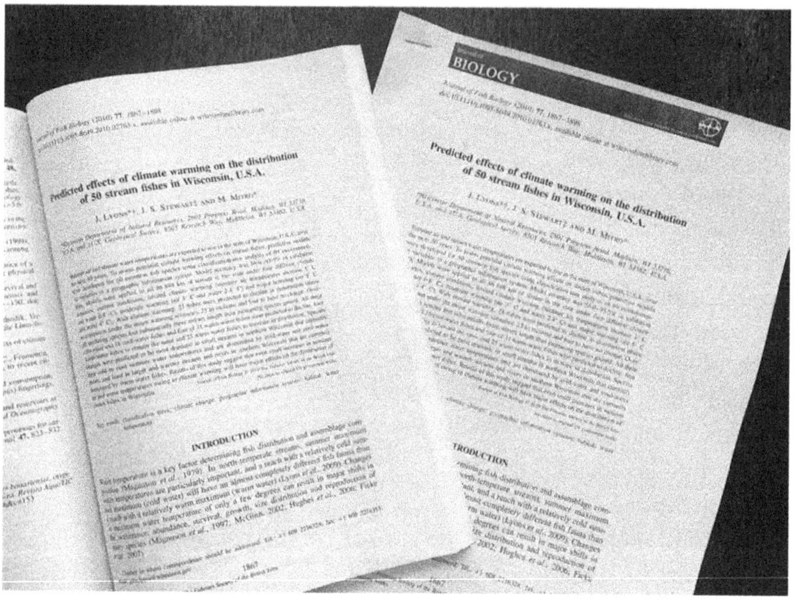

Figure 6.3

Navigating a Journal's Website

Many (but not all) journals are presented on a website provided by the journal's publisher. So, to find a journal on the internet, run a Google search on the title of the journal. It may help to put the title of the journal within quotation marks to search it as a phrase, and you may also want to add the word "journal" to your search if the word "journal" is not already in the title. Here are two example searches:

- **"Journal of Clinical Oncology"**
- **"Astrophysics and Space Science" journal**

Assuming you can find the journal's homepage, it usually has navigation that allows you to browse from volumes/issues to individual articles, like you see on the left in Figure 6.4.

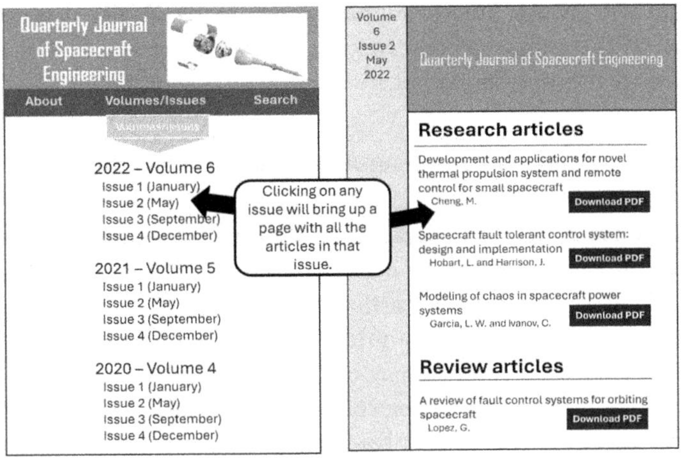

Figure 6.4

From that page, you can click on a link to an issue, such as Issue 2 from Volume 6. This will pull up a page showing all the articles published in that issue (shown on the right in the image above). From there, let's say you decide to click on the title of that first research article: *Development and applications for novel thermal propulsion system and remote control for small spacecraft.*

That will lead you to the article's webpage (shown in Figure 6.5), which typically contains an **HTML version** of the article and a link to the **PDF version** of the article. That

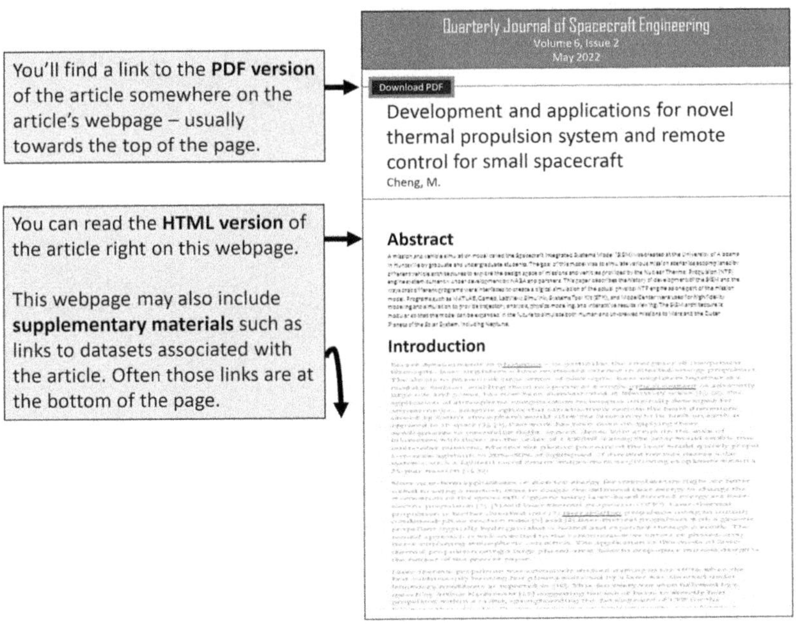

Figure 6.5

webpage might also provide links to **supplementary materials** like data sets associated with the article. (There may be other links and information as well, but the elements just mentioned are the most important.)

Making the Connection between an Article and Its Journal

Now suppose you're conducting research on a topic and you come across a relevant journal article in a library database or while searching the internet. You decide to download the PDF to read the entire article. Because you are encountering this article PDF all on its own, and not within a print journal issue that shows the name of the journal on the issue's front cover, it may not be clear what journal the article belongs to. It's just as if you encountered a book chapter that had been torn out of a book—you wouldn't know what book that chapter belongs to.

So, how do you determine what journal your article was published in? Look on the first page of the article. The name of the journal is usually printed at the very top or bottom of the page. Sometimes the journal name is in small font and sometimes it's abbreviated. In Figure 6.6, you can see the article is published in the journal *Landscape Ecol.*, which is an abbreviation of *Landscape Ecology.*

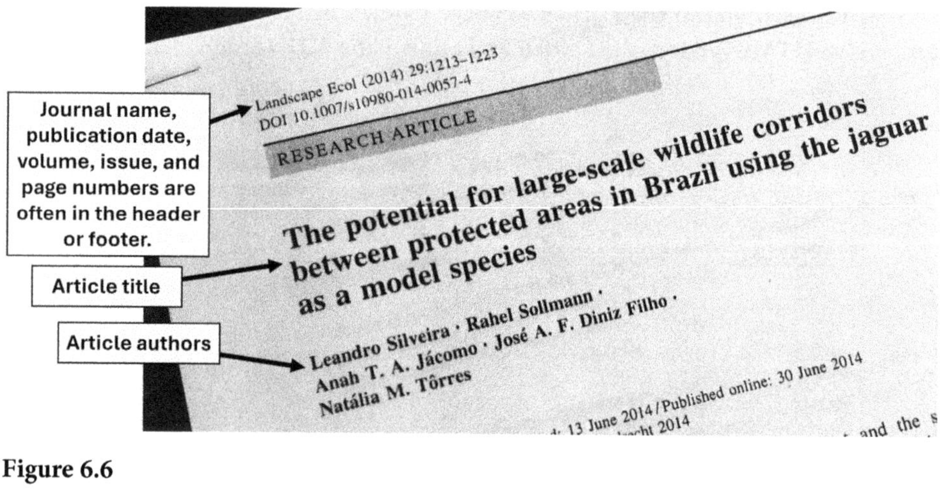

Figure 6.6

Don't forget that depending on the journal, all this information may be at the *bottom of the page*—in the footer—instead of at the top. Also notice that next to the title of the journal you can usually find the volume number, issue number, and page numbers for the article. So, when you need to *cite* an article, you can usually find all the elements you need on the first page.

Different Names for Journals and Journal Articles

It's fairly common to hear people refer to journals by any of these names:

- **scholarly journals**
- **academic journals**
- **peer-reviewed journals**
- **refereed journals**

When people say "scholarly journals" they're talking about journals that contain articles that are scholarly in nature, meaning the articles are written by authors who are expert scholars in their discipline. Often these expert scholars work at academic institutions like colleges and universities, which is where the name "academic journals" comes from. In fact, the labels "scholarly" and "academic" are just two different ways of saying the same thing. Consider them to be synonyms. The labels "peer-reviewed" and "refereed" are also synonyms. Those labels refer to journals in which the articles must pass through a quality control process—called "peer review"—before being published. (You'll learn more about peer review in Chapter 14.)

The thing is, most people would define a "journal" as a type of publication that has a peer-review process in place, and contains articles written by expert scholars who typically work at academic institutions. Thus, there's no meaningful distinction among any of those four ways of referring to journals that you see listed above; just consider all of them to be synonyms of the word "journal."

People refer to journal *articles* in a few different ways too, so don't let that trip you up. Besides being called **articles**, they're sometimes called:

- **papers**
- **technical papers** (especially in engineering)
- **studies**

And collectively journal articles might be referred to as **the journal literature** or **the scientific literature.**

Identifying Predatory Journals

Let's talk about how to identify **predatory journals**, which are not legitimate journals. Predatory journals are published by **predatory publishers**, which you learned about in Chapter 2. The easiest way to check whether a journal is predatory is to run a Google search on **predatory journal lists**. You'll get results that lead you to lists of predatory journals and then you can check those lists to see if a particular journal turns up on any of them. You can also try entering the title of a journal into a Google search along with the word "predatory" to see what information you can find.

Finding Journal Articles

If you need a specific journal article or if you're browsing for journal articles on a topic, then your campus library will offer several databases that can help. If you like, you can start by searching your library's **discovery tool** (which you'll learn more about in Chapter 20). If your library offers this kind of database, it'll likely be made available as a search box on your library's home page that invites you to search articles, books, media, and more—or has wording similar to that. For example, you can use your library's discovery tool to look for a specific journal article by entering its title, just like you learned to do to find a specific book in Figure 3.1. Alternatively, you can enter keywords to look for journal articles on a topic. Once you've run your search, look for filters (which are usually to the left of your results) to limit your results to journal articles only—or to peer-reviewed content only.

If you don't see a search box like this on your library's homepage, or if the search you conducted didn't find what you needed, then be aware that your library probably offers many other databases for finding journal articles. In fact, if you're conducting serious research on a topic and you need to find journal articles, then discovery tools are often not the best database to use (although they are useful when you need to track down a particular article). So, you should also explore your library's **article databases** (such as Web of Science, PubMed, Scopus, Compendex, etc.) to search for journal articles. To find the article databases your library has to offer, look on your library's homepage for a link with any of these names, or a similar sounding name:

- **Databases**
- **All Databases**
- **Databases A–Z**
- **Find Databases**
- **Research Databases**
- **See All Databases**

In Figure 6.7 you can see an example—on this library's homepage you'd click the link called "databases."

Once you click that link, you'll see options to navigate to databases by title, and you may also see options to navigate to the best databases for a *particular subject area*. So, if your topic relates to biology, look for an option to view the best databases for biology

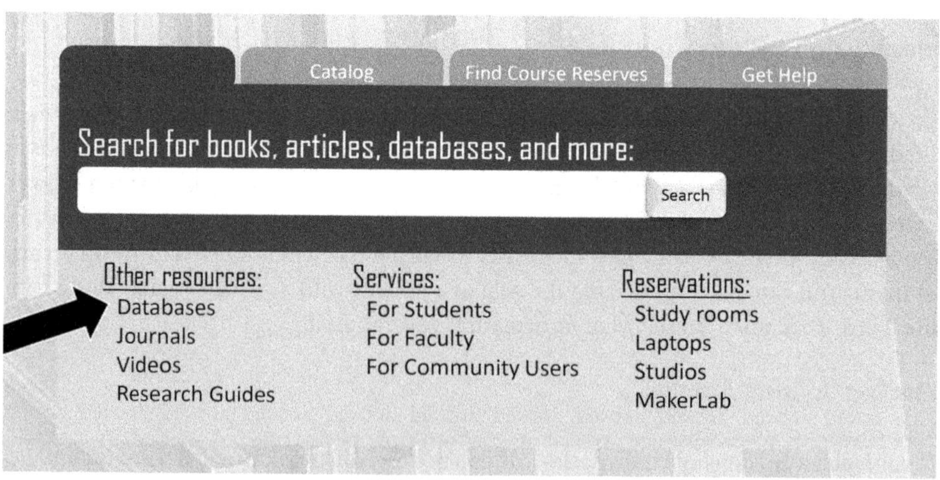

Figure 6.7

topics. You'll learn more about article databases and their coverage in Chapter 20 and you'll learn how to choose the best ones to search in Chapter 23.

Don't forget you can also search Google Scholar for a particular article, or you can enter keywords into Google Scholar to search for journal articles on a topic. You can also search Google for the title of a particular journal article to see if you can find a copy for free on the internet. Google can usually locate freely available journal articles that have been uploaded to sites like **ResearchGate** or **Academia.edu**, which are social networking sites for researchers. If you explore all these options and still can't find what you need, then ask for help or consult the section below.

Dealing with Access Problems

If there's a journal article you want to read, but your library doesn't seem to have it, nor can you find it for free on the internet, then request it through your library's **document delivery/interlibrary loan** service. You learned about this service in previous chapters, but as a reminder, this service can supply you with sources your library doesn't have. You should be able to find a link to this service on your campus library's homepage, but if you need help finding and using it, ask library staff for assistance.

Distinguishing Journal Articles from Other Sources

A print journal issue looks somewhat like a print magazine issue but the cover is usually more serious looking and less flashy. Plus, the articles contained within will be dense and technical, and you won't see advertisements. Distinguishing journal articles from other sources in an *online* environment can be harder, so let's go over some strategies for doing that.

Suppose you're searching a library database and you aren't sure if it's presenting you with results that are journal articles. As you've learned in previous chapters, you can look for filters to refine your results to *just* journal articles, or you can check to see if the database identifies journal articles in your results. Take another look at the database results shown in Figure 3.7 and you'll see that the second one is labeled as a scholarly journal article.

These labels make it clear what type of source you're viewing, but as you know, not all databases label results by source type, and most notably, Google doesn't. So, let's revisit that Google search on **coral reef conservation** from previous chapters. If you recall, it turned up a newspaper article and a magazine article, but that search retrieved a journal article as well (see Figure 6.8).

How would you know that last result is a journal article? As you learned in Chapter 2, it takes practice and experience to pick up on source clues within your Google results. For this particular result, the clues indicating it's a journal article are as follows:

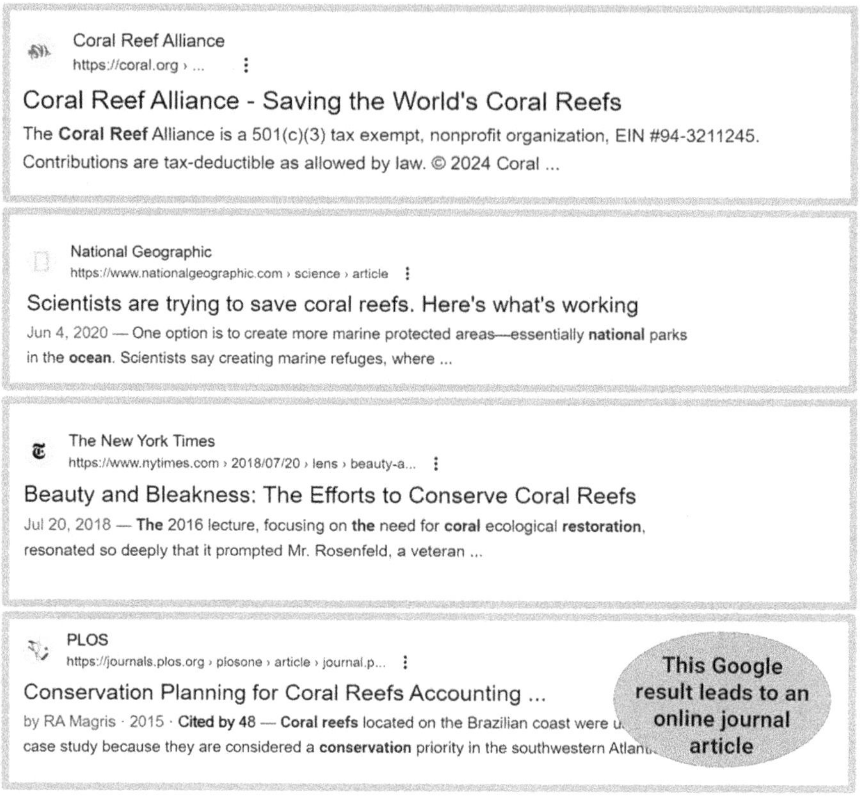

Figure 6.8

- The result is coming from *PLoS*, which is a well-known open-access journal publisher (*PLoS* stands for Public Library of Science).
- The article has been "Cited by 48," which is a metric associated with journal articles.

As you've learned in previous chapters, once you click on a Google result and view the source, you can look for more clues about what type of source you're viewing (see Figure 6.9). To distinguish journal articles from other types of sources, look for these characteristics:

- Note whether the article title sounds like it describes research.
- Look for dense, technical writing.
- See if you can find the journal name, volume number, issue number, page numbers, and/or a DOI number somewhere in the header or footer of the first page of the article PDF.
- Page through the article to see if the text is organized into sections that include an introduction, methods, results, and discussion. (Not all journal articles have these sections, but many do.)

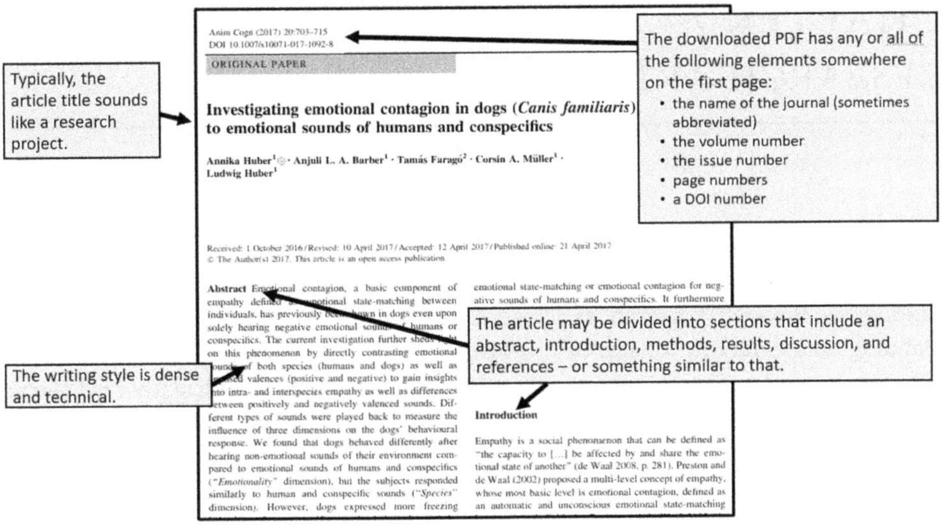

Figure 6.9

To sum up, what should science and engineering students know about journals?

1. Journals are published in installments called issues. Each issue contains several articles. A whole year of issues is called a volume.
2. Most journal articles describe new discoveries, applications, or inventions that resulted from research, and their purpose is to communicate these findings to other researchers.
3. Nowadays, most journal articles are accessed in PDF form, rather than within a print journal issue.
4. You can usually find all the details you need to construct a journal article citation (including journal name, date, volume, issue, and pages) on the first page of the article PDF.
5. Journal articles can be found on the internet, from library databases, as well as through library services such as document delivery/interlibrary loan. Libraries provide staff to help you locate journal articles when you need assistance.

What's Next?

In the next chapter you'll learn about a specific type of journal article called a **research article**.

CHAPTER 7
RESEARCH ARTICLES

One of the main ways that scientists and engineers publish and share their research findings is through research articles. You must understand this source if you're planning on a career in science or engineering. In this chapter you'll learn about the purpose, content, and structure of research articles, and how to efficiently read them. This chapter will also introduce you to different research study designs.

An Overview of Research Articles

The purpose of a **research article** is to describe a research project that's been conducted and to record the findings of that research for posterity. Research articles usually report on the results of a *single* study that was conducted. A review article (which you'll learn about in the next chapter) is different; it provides a summary of findings from multiple, related research articles.

Because research articles are one of two main sources (the other being **conference papers**) where research discoveries are documented and shared, they are a very important source for scientists and engineers. In your future career, you'll find plenty of reasons to regularly seek out and refer to research articles. For example, you'll need to consult them in order to understand which research projects have, or have not, already been conducted. You'll need to refer to previously published research articles to set a starting point for embarking on new research projects *and* to provide context for your research projects. Consulting research articles will also provide you with ideas for methods, equipment, and techniques to apply to your own projects. As research articles formally document who is working on what, they'll also help you identify potential collaborators or experts to consult.

Different Ways People Refer to Research Articles

Before delving into the characteristics of research articles, let's talk about the different ways people refer them. Sometimes you'll hear them called **empirical articles**, because they describe research that's based on observation and data-gathering. You'll also hear them called **original articles**, **primary articles**, or **primary research articles**. That's because in science and engineering, research articles have qualities that make them **primary sources**. You'll learn more about primary sources Chapter 16, but for now think of a primary source as any source that documents original, brand-new information.

Because research articles make up the bulk of all journal articles, you'll hear people refer to them simply as **journal articles**, even though they're actually a specific type of journal article. So, if a college instructor were to say to you "Go find some journal articles on your topic" they probably mean: "Go find some research articles on your topic." Also, don't forget in the previous chapter you learned that journal articles are sometimes called **papers** or **studies**. When you hear these terms, the person using them is usually referring specifically to research articles.

The Structure of Research Articles

Most research articles are laid out according to a standard structure that acts as a prompt for authors to document the methods they used and the results they gathered, as well as provide a discussion about what the results signify. That structure is as follows:

- **Title**—Article titles usually provide a very concise overview of what the article is about.
- **Authors**—The article's authors are usually listed just under the title, often with information about their institutional affiliations, and usually with the email provided for at least one author who is called the **corresponding author**. (The corresponding author is the author you would correspond with if you had a question about the article.)
- **Abstract**—This is a short summary of the entire research project.
- **Introduction**—In this section the authors explain what research they set out to conduct and why.
- **Methods**—This is where the authors describe the procedures and equipment they used to conduct their research project.

- **Results**—The authors use this section to display the data they collected, often using graphs, figures, or tables.
- **Discussion**—In this section, the authors explain what they learned from their study, including the significance of it, and how it advances our knowledge of the world.
- **References**—This is where the authors list citations to the publications they referred to throughout their article.

Figure 7.1 shows the first page of a research article. Notice this page shows the first three elements of a typical research article's structure: the title, authors, abstract, and introduction. On the pages that follow you'd find sections describing the methods, results, discussion, and references.

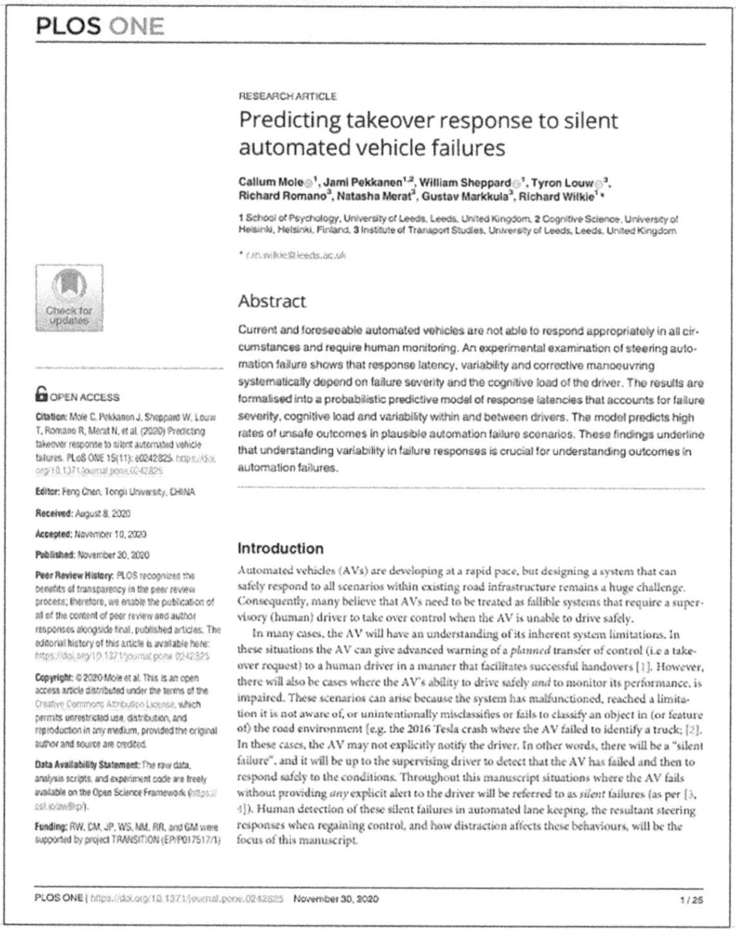

PLOS ONE

RESEARCH ARTICLE

Predicting takeover response to silent automated vehicle failures

Callum Mole[1], Jami Pekkanen[1,2], William Sheppard[1], Tyron Louw[3], Richard Romano[3], Natasha Merat[3], Gustav Markkula[3], Richard Wilkie[1]*

1 School of Psychology, University of Leeds, Leeds, United Kingdom, 2 Cognitive Science, University of Helsinki, Helsinki, Finland, 3 Institute of Transport Studies, University of Leeds, Leeds, United Kingdom

* r.m.wilkie@leeds.ac.uk

Check for updates

OPEN ACCESS

Citation: Mole C, Pekkanen J, Sheppard W, Louw T, Romano R, Merat N, et al. (2020) Predicting takeover response to silent automated vehicle failures. PLoS ONE 15(11): e0242825. https://doi.org/10.1371/journal.pone.0242825

Editor: Feng Chen, Tongji University, CHINA

Received: August 8, 2020

Accepted: November 10, 2020

Published: November 30, 2020

Peer Review History: PLOS recognizes the benefits of transparency in the peer review process; therefore, we enable the publication of all of the content of peer review and author responses alongside final, published articles. The editorial history of this article is available here: https://doi.org/10.1371/journal.pone.0242825

Copyright: © 2020 Mole et al. This is an open access article distributed under the terms of the Creative Commons Attribution License, which permits unrestricted use, distribution, and reproduction in any medium, provided the original author and source are credited.

Data Availability Statement: The raw data, analysis scripts, and experiment code are freely available on the Open Science Framework (https://osf.io/aw8kp/).

Funding: RW, CM, JP, WS, NM, RR, and GM were supported by project TRANSITION (EP/P017517/1)

Abstract

Current and foreseeable automated vehicles are not able to respond appropriately in all circumstances and require human monitoring. An experimental examination of steering automation failure shows that response latency, variability and corrective manoeuvring systematically depend on failure severity and the cognitive load of the driver. The results are formalised into a probabilistic predictive model of response latencies that accounts for failure severity, cognitive load and variability within and between drivers. The model predicts high rates of unsafe outcomes in plausible automation failure scenarios. These findings underline that understanding variability in failure responses is crucial for understanding outcomes in automation failures.

Introduction

Automated vehicles (AVs) are developing at a rapid pace, but designing a system that can safely respond to all scenarios within existing road infrastructure remains a huge challenge. Consequently, many believe that AVs need to be treated as fallible systems that require a supervisory (human) driver to take over control when the AV is unable to drive safely.

In many cases, the AV will have an understanding of its inherent system limitations. In these situations the AV can give advanced warning of a *planned* transfer of control (i.e a takeover request) to a human driver in a manner that facilitates successful handovers [1]. However, there will also be cases where the AV's ability to drive safely *and* to monitor its performance, is impaired. These scenarios can arise because the system has malfunctioned, reached a limitation it is not aware of, or unintentionally misclassifies or fails to classify an object in (or feature of) the road environment (e.g. the 2016 Tesla crash where the AV failed to identify a truck; [2]. In these cases, the AV may not explicitly notify the driver. In other words, there will be a "silent failure", and it will be up to the supervising driver to detect that the AV has failed and then to respond safely to the conditions. Throughout this manuscript situations where the AV fails without providing *any* explicit alert to the driver will be referred to as *silent* failures (as per [3, 4]). Human detection of these silent failures in automated lane keeping, the resultant steering responses when regaining control, and how distraction affects these behaviours, will be the focus of this manuscript.

Figure 7.1

This organizational structure is sometimes referred to as the **IMRaD structure**, which stands for **I**ntroduction, **M**ethods, **R**esults, **a**nd **D**iscussion. While this is the most commonly used organizational structure for research articles, it's not the only structure used. For instance, in some research articles you might find these sections in a different order, while other research articles might add a "conclusion" section after the "discussion." Research articles in engineering and computer science tend to have even more variation. The article below offers a good example:

> Zanini, M. A., et al. (2017). Probabilistic seismic risk forecasting of aging bridge networks. *Engineering Structures, 136*: 219–232.

Below you see an outline of that article's contents. Notice that it does not have special sections devoted to methods or results:

- Title
- Abstract
- Introduction
- Improving classical EAL methodology estimation
- Time-dependent seismic hazard
- Time-dependent seismic fragility
- Time-dependent exposure
- Loss assessment and optimal retrofit planning frameworks
- Application of framework to a portfolio of deteriorating bridges
- Conclusions

In general, articles that use mathematical calculations to solve problems tend to veer the farthest from the IMRaD structure. For these articles, the math itself is the method and the math creates a result, so it's not always necessary for the author to discuss methods and results at length.

Strategies for Reading Research Articles

The structure of a research article allows readers to skip around and read the sections that interest them the most, as you can see from Figure 7.2. This is extremely useful because

	I'm Looking For . . .							
	The really big picture	Main details about the project	Nitty gritty details about the project	Evidence that was gathered	Methods and data analysis	Important findings and implications	Limitations	How this research connect to previous research
Title	X							
Abstract	X							
Introduction		X						X
Methods			X		X			
Results			X	X				
Discussion/ Conclusions		X				X	X	
References								X

Figure 7.2

reading a research article from start to finish is a time commitment and researchers are always looking for ways to save time!

Also, it's common for researchers to read certain parts of an article first to determine whether it's worth their time to read the article in its entirety. Typically, they'd first look at the **article title** to determine whether the article sounded useful for their needs. If so, they'd probably go on to read the **abstract**. If the article still sounded relevant after that, they might jump to the **discussion** to read what the authors learned from their research project. Or they might skip to the **methods** section to see what materials and techniques the authors used. Even if they read the entire article, they may not read each section in order.

Of course, as a student, you may encounter assignments where you're asked to read articles in their entirety, from start to finish. Giving individual articles a really close read can help you get a much better feel for the content, writing style, and purpose of journal articles—so dive in!

When you get more familiar with journal articles, you may notice that certain sections have their own internal structure. For example, the **introduction** usually starts off with the authors discussing the broad context of their research project and how it relates to and builds upon what we already know from previously published research. The authors might then move on to discuss gaps in our knowledge of that research area. Finally, at the end of the introduction, they'll discuss the goals of *their* research project and the specific research question they aim to tackle. So, the introduction starts off general and then gets specific. Let's take a look at some excerpts from the introduction of this article:

> Ware, I. M., et al. (2014). Conflicting management objectives on the Colorado Plateau: understanding the effects of bison and cattle grazing on plant community composition. *Journal for Nature Conservation*, *22*(4): 293–301.

The authors begin their introduction by providing readers with the broad context for their research project:

> Rangeland ecosystems provide ecosystem services including food and fiber production, biodiversity, climate regulation, water supply, and recreation. However, trade-offs among ecosystem services exist, meaning that a particular land management practice may promote one service but reduce another, leading to conflict (Bennett et al., 2009).

Then, you'd see more paragraphs of text laying out additional background, including how the research project relates to and builds upon previously published research. Of course, the authors would cite all that previously published research, so it's typical to see lots of references to other sources in the introduction. Finally, you'd get to the last paragraph of the introduction where the authors narrow in on their specific goals and research questions:

> Our goal was to inform future management decisions by determining the extent to which bison have altered plant communities on cattle winter range in the

> Henry Mountains. We focused on three research questions: (1) Is the presence of bison increasing grazing intensity or altering the season of utilisation on these rangelands? (2) Do areas grazed by bison and cattle, areas grazed only by cattle, and ungrazed areas differ in species composition? If so, do current differences in plant community composition provide evidence that bison grazing is causing degradation? (3) Based on long-term remotely sensed vegetation data, does the timing of vegetation change coincide with the increase in bison use?

So, if you want to know why researchers undertook a research project and how their project connects with previous research that's been conducted, you'd read the first part of the introduction. If you want to know what specific project the researchers embarked upon, you'd skip to the end of the introduction.

The Narrow Scope of Research Articles

In the section above you read the last paragraph from the introduction of a research article and you learned that this is where you can expect authors to explicitly state the research questions they intend to tackle. But did you notice how *specific* the research questions were in that excerpt? That's typical. Research projects tend to be very narrow and focused because you simply can't do meaningful, productive research on broad questions. Suppose the authors of that article tried to investigate a much broader question such as "How should we manage plant communities on the Colorado Plateau?" It would be impossible to set up a study that would answer such a broad question because there are so many different variables involved, such as geography, climate, altitude, concentration of grazing animals, land use by humans, etc. Instead, researchers must focus on narrow problems that can be appropriately controlled and tested *and* that will produce results that can be independently verified.

Now that you know how narrowly focused research articles are, consider how inappropriate it would be to seek them out when you're in need of general information or answers to broad questions. Instead, look for that sort of information in a book since books offer comprehensive summaries of information, rather than nitty-gritty details. **Review articles** (which you'll learn about in the next chapter) can also be useful when you're seeking an overview or summary of information. Remember, whenever you can match your information need to the appropriate source, you'll save yourself time and effort.

Research Methods and Study Designs

In some disciplines—such as health sciences, psychology, and sociology—it's common to categorize research articles by the type of data gathered and how they were gathered. For example, an article might describe an **experimental study** or an **observational study**, or it might use **quantitative research methods** or **qualitative research methods**.

Quantitative research involves collecting and analyzing numerical (quantifiable) data in the form of measurements, statistics, percentages, scores, etc., to conduct empirical research. Quantitative research methods are used across the sciences, social sciences, and engineering.

Qualitative research involves collecting non-numerical data in the form of interviews, surveys, focus groups, etc., in order to understand human experiences and behaviors. These methods are more commonly used in the social sciences, but not exclusively. Below is an example of a qualitative study:

> Sharif, F. & Masoumi, S. (2005). A qualitative study of nursing student experiences of clinical practice. *BMC Nursing, 4*(1): 1–7.

Some research articles can be further categorized by the type of research methodology used, called the **study design**. For example, in health sciences, a **clinical trial** is a type of study design in which human subjects are given a prospective treatment therapy and then compared to another set of human subjects who were not given that therapy, in order to determine how effective the therapy is. A more rigorous version of this study design is called the **blinded, randomized, controlled trial**. In this version, human subjects are randomly assigned to test the therapy and neither the subjects nor the researchers know who is getting the therapy and who isn't. This reduces bias and placebo effects and provides more conclusive results.

Additional examples of study designs in the health sciences include **non-randomized controlled trials**, **crossover designs**, **cohort studies**, **case-control studies**, and **cross-sectional surveys**. Depending on your chosen discipline you may need to learn a lot more about particular research methods and study designs, but there are many books and websites available to help you understand the distinctions.

Research Articles Are Skewed towards Reporting Positive Results

Research articles that make it to publication are skewed towards those that describe successful research projects. A successful research project is one where the researchers had a hypothesis, tested it, and got statistically significant results that supported their hypothesis. When results support a hypothesis, they're called **positive results**. If, on the other hand, a research project fails to return results that support a hypothesis, we call those **negative results**.

Most journal publishers aren't interested in publishing studies that produced negative results. They're concerned those articles won't get cited very much, thus lowering the impact factor of the journal. (An **impact factor** is a metric indicating how useful and prestigious the journal is; you'll learn more about this metric in Appendix B.) So, journal publishers tend to reject studies showing negative results.

This is a problem because negative results are important to communicate to others. As the author of the article shown below explains, if you don't share negative results then "other scientists can't learn from them and end up repeating failed experiments, leading to a waste of public funds and a delay in genuine progress."

Mehta, D. (2019). Highlight negative results to improve science. *Nature*. DOI: https://doi.org/10.1038/d41586-019-02960-3

When journal publishers reject studies showing negative results, this has adverse effects on researchers as well. To understand why, let's take a few steps back. In Chapter 2 when we talked about predatory publishers, you learned that many employers—and especially academic employers—regard publishing as a measure of their researchers' productivity. From the employer's perspective, publishing very few articles is equivalent to being unproductive and could lead to firing the researcher. This is where the saying "publish or perish" comes from.

Now consider that over the course of a researcher's career, lots of their hypotheses will be wrong and will fail to produce positive results—that's just how science and engineering slowly chug along with two steps forward and one step back. But when researchers can't publish their negative results, they'll appear to be unproductive to their employer, even though the researcher has been working hard. Furthermore, a publishing system "that disfavours negative results not only distorts the scientific literature directly, but might also discourage high-risk projects and pressure scientists to fabricate and falsify their data" according to Daniele Fanelli, who authored the paper below:

Fanelli, D. (2012). Negative results are disappearing from most disciplines and countries. *Scientometrics*, 90(3): 891–904.

So, pressure on researchers to publish positive results (or lose their jobs!) can act as an incentive for them to tweak, exaggerate, or even falsify data. The pressure to publish or perish also means that researchers may take liberties with their study design, watering it down to make it less robust if that will make their results more statistically significant and exciting—and thus more likely to get published.

Are there any attempts to fix the problems outlined above? Yes. The most promising initiative is the "Registered Reports" program from the Center for Open Science (COS). In this program, COS negotiated partnerships with several hundred journals to allow researchers to publish negative results under certain circumstances. Specifically, as long as the researcher has designed their study well (judged through an initial peer-review process) and they follow through with their study exactly as they designed it (judged through a second peer-review process) then their research is guaranteed to be published even if it generates negative results.

So, just be aware that the journal article publishing industry has some imperfections (and you'll learn about more problems in later chapters), but there are initiatives to make it better.

To sum up, what should science and engineering students know about research articles?

1. Research articles are a type of journal article.
2. Research articles describe a research project that was conducted and the findings of that project.
3. Most research articles are organized into sections that include an abstract, introduction, methods, results, and discussion—or variations on that theme.
4. The fact that research articles are organized into sections allows you to read them strategically, zeroing in on information that'll be the most useful to you.
5. The research described in research articles is usually very narrow in scope.
6. Many research articles can be further categorized by the types of research methods used, or by whether they follow a specific study design.
7. The research articles that make it to publication are skewed towards those that report on positive results.

What's Next?

In the next chapter you'll learn about another type of journal article called a **review article**. Review articles are extremely useful and serve a completely different purpose than research articles.

CHAPTER 8
REVIEW ARTICLES

Review articles help researchers reflect and take stock of research progress. These articles help us understand what we already know about a research area and how confident we can be in that knowledge. These articles also point out gaps in our knowledge that need to be filled. This chapter will explain different types of review articles, the purposes they serve, and how they are used by researchers in science and engineering.

An Overview of Review Articles

A **review article** provides a review of the current state of knowledge of a topic—or in other words, it provides an overview and summary of progress in a particular area of research. So, the purpose of a review article is to help readers quickly and easily understand what progress has been made in a particular area of research.

When authors write review articles, their first step is to search for journal articles about that area of research. This process is called conducting a **literature search**. Typically, most of the articles gathered are **research articles**, which you learned about in the last chapter. But other types of articles, such as review articles or perspective articles, might also be examined. Next, they read those articles and assimilate their contents. They'll look at the history of progress in that area of research, as well as any patterns, conclusions, overlaps, discrepancies, etc., that emerge from the articles as a group. Finally, they'll write up what they learned to provide the reader with a summary of what we currently know about that research area. Many review articles also identify gaps in our knowledge; pointing out gaps helps other researchers understand where future research efforts should be directed. Here is an example of a review article:

> Albalawi, E. K. & Kumar, L. (2013). Using remote sensing technology to detect, model and map desertification: a review. *Journal of Food, Agriculture and Environment, 11*(2): 791–797.

To create this **review article,** the authors searched for and gathered other journal articles—primarily **research articles**—where remote sensing technology was used to model and map desertification. Then they summarized the findings of those articles to produce an overview of how remote sensing is being used to map desertification.

A review article is considered to be a **secondary source**, unlike research articles, which are **primary sources**. Secondary sources don't document the results of brand-new, original research. Instead, they take pre-existing information from sources like research articles, and then summarize, synthesize, and/or analyze that information. You'll learn more about primary and secondary sources in Chapter 16, but for now, the main takeaway is that review articles summarize the findings of multiple research articles. Keeping that in mind, suppose you wanted to get yourself up to speed on an area of research that you're not that familiar with. Can you see how helpful and convenient it would be to read a review article on that topic?

The Structure of a Review Article

As you now know, most *research* articles are organized in a standard structure that includes an abstract, introduction, methods, results, and discussion. Review articles sometimes have the same structure and sometimes they don't. Here's an example of a review article that uses the same structure as a research article:

> Hegde, J. & Rokseth, B. (2020). Applications of machine learning methods for engineering risk assessment—a review. *Safety Science, 122*: 104492.

Below you see how that article is organized; note the sections are the same as those you'd expect in a research article:

- **Title:** Applications of machine learning methods for engineering risk assessment—a review
- **Abstract**
- **Introduction**
- **Methods**
    - **Literature survey process**
    - **Literature classification**
- **Results**
    - **Relevant literature**
    - **Results obtained from citation information**
    - **Results obtained from reviewing the articles**
- **Discussion**
- **Conclusion**
- **References**

Are you wondering why a review article would have a methods section? After all, the authors aren't conducting an *experiment* where they need to explain the methods they used. However, the authors *are* using particular search strategies to gather articles for their review, and so they might explain those strategies in the methods section of their review article.

In the outline for the review article shown above, you can see that's exactly what these authors did. Their methods section explains that they searched the databases Scopus and Engineering Village, and they show some of the keywords searches they conducted such as:

- Machine Learning AND Engineering AND (Risk OR Safety)
- Artificial Intelligence AND Engineering AND (Risk OR Safety)
- Data Mining AND Engineering AND (Risk OR Safety)

However, not all review articles are structured the same way, so now let's look at a review article that does *not* have the same structure as a research article:

Bouffard, D. & Wüest, A. (2019). Convection in lakes. *Annual Review of Fluid Mechanics, 51*: 189–215.

Below you see the different sections of that article. Note that it doesn't include a methods, results, or discussion section. Instead, it's organized into sections that directly relate to the topic of the article:

- **Title**: Convection in lakes
- **Abstract**
- **Introduction**
- **Convection induced by boundary fluxes**
- **Shear-induced convection over slopes**
- **Thermobaric convection in deep lakes**
- **Double-diffusive convection**
- **Bioconvection in lakes**
- **Conclusions and future directions**
- **Future issues**
- **Literature cited**

So, in some review articles, the authors don't outline their methods for gathering articles and instead just organize the article as they see fit and create their own section headings.

Remember that it's common for the authors of a review article to point out gaps in our knowledge and where research efforts should be directed next. In this article, if you wanted to quickly zero in on that sort of information you could skip to the sections on **Conclusions and future directions** and **Future issues**. Not all review articles have sections that so clearly indicate where this sort of content will be found, but this one does.

Types of Review Articles

Whereas research articles differ in the **study designs** used, review articles also vary by type. For example, there are traditional **literature reviews**, **state-of-the art reviews**, **perspective articles**, **systematic reviews**, and **meta-analyses**. These five types of review articles are fairly common, but in some disciplines, such as the medical sciences, you'll regularly encounter more types of review articles, such as **scoping reviews**, **rapid reviews**, **mapping reviews**, etc. We'll only cover the most common types here, but if you want to learn more, just do a Google search on **review article types** or search Google for a particular type of review article, such as a **mapping review article**, if you want to learn more about that type.

A **traditional literature review**, also called a **narrative review**, is what you've been reading about so far. That is, it's a type of article that provides an overview and summary of the published literature in a specific area of research.

A **state-of-the-art review** is like a traditional literature review except that it only examines the very latest developments and advancements in an area of research. These reviews provide a benchmark of recent progress, and they may discuss emerging trends as well. State-of-the-art reviews are fairly common in engineering. Here's a good example:

> Hutmacher, D. W. (2001). Scaffold design and fabrication technologies for engineering tissues—state of the art and future perspectives. *Journal of Biomaterials Science, Polymer Edition*, *12*(1): 107–124.

This is a good place to mention that you should be aware of the publication date of a review article—and especially a state-of-the-art review article. If it's old, then it won't include the most recent developments. Obviously, a five-year-old state-of-the-art review—or older—is no longer going to be state-of-the-art.

A **perspective article** shares some of the same characteristics as a narrative review and a state-of-the-art review except these articles offer a personal perspective on an area of research. For example, authors may present their vision for where future research efforts should be directed. Or, they may propose solutions to controversies within an area of research, or express their opinions on how to resolve controversies. In general, these articles are more speculative and opinionated, but the authors still base their ideas on a close examination of existing research.

A **systematic review article** is more formal and focused. In a systematic review article, the authors collect information relevant to a specific research question. They'll comprehensively search for research articles that have produced results and evidence relevant to that research question. Then, they'll examine those articles for the quality and robustness of their research methods, excluding any that don't meet certain standards. (For example, articles might get excluded if they don't have large sample sizes, or statistically significant results, or weren't conducted using a double-blind study design.) For the articles that remain, the authors will examine each, looking carefully at the evidence presented in the research results. Their objective is to determine whether the research evidence from

all those articles is converging on the same answer to the research question. If so, this collective evidence raises our level of confidence that we have the right answer to that question. So, a systematic review article helps us understand how confident we can be that we've resolved a research problem.

A **meta-analysis** is much like a systematic review article except the focus is on gathering *data* from multiple, high-quality, research articles to answer a particular research question. The authors of the meta-analysis will then re-crunch those data to arrive at an answer to the research question. The answer they arrive at will have more statistical clout, because it's drawn from more extensive data; thus, it should provide a more conclusive answer to the research question.

Both systematic reviews and meta-analyses interpret evidence gathered from multiple, robust research articles. In doing so, they provide stronger evidence that a research question has been resolved than any of those research articles could do on their own. Thus, systematic reviews and meta-analyses have a higher level of authority than other types of articles, and for this reason they are often used to provide guidance in making policy decisions and developing best practices. In fact, these two types of review articles were first developed and used in the discipline of medicine for just that purpose: to collect the best and strongest evidence about which treatments and therapies should be used on patients. If you're planning on a career in the medical sciences, you'll probably see infographics (such as Figure 8.1) that rank medical information sources according to our confidence in the accuracy of their conclusions.

The use of systematic reviews and meta-analyses now extends far beyond establishing best medical treatments and practices, plus these types of articles have been widely adopted in disciplines outside the medical sciences. Table 8.1 shows a couple of examples of systematic reviews and meta-analyses, one from forestry and one from sports

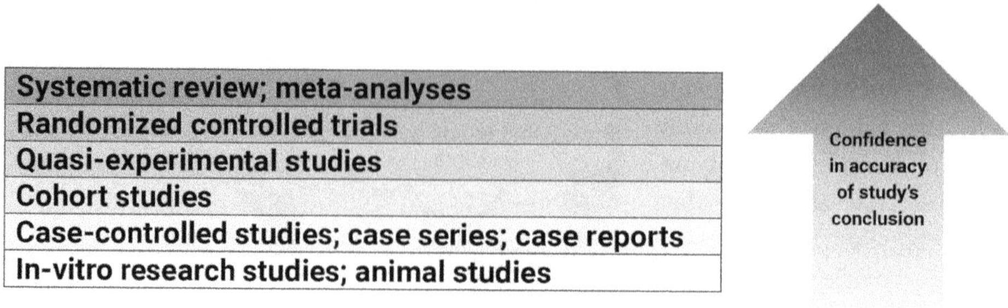

Figure 8.1

Table 8.1

Article	Fulé, P. Z., et al. (2012). Do thinning and/or burning treatments in western USA ponderosa or Jeffrey pine-dominated forests help restore natural fire behavior? *Forest Ecology and Management*, *269*: 68–81.	Hallgren, M., et al. (2017). Exercise as treatment for alcohol use disorders: systematic review and meta-analysis. *British Journal of Sports Medicine*, *51*(14): 1058–1064.
Research question	Do thinning and/or burning treatments on ponderosa pine and related forests in western USA produce restoration of natural fire behavior?	What are the effects of exercise for people with alcohol use disorders (AUDs) across multiple health outcomes?
Research articles examined	139 articles met the inclusion criteria	21 articles met the inclusion criteria
Conclusion	"Overall, however, meta-analysis of the literature to date strongly indicates that thinning and/or burning treatments do have effects consistent with the restoration of low-severity fire behavior."	"Available evidence indicates exercise appears not to reduce alcohol consumption, but has significant improvements in other health outcomes, including depression and physical fitness."

medicine. Notice the research question that was examined, the number of articles that were included in the analysis after the authors excluded articles that didn't qualify, and the conclusion to the question.

There are a couple of additional things to note about the two examples shown in the table above. First, you can't always expect the article title to explicitly indicate it's a systematic review or meta-analysis. Second, some authors combine both a systematic review and meta-analysis in one article.

Where Are Review Articles Published?

Review articles are published in journals just like research articles. Some journals contain a mix of research articles and review articles in every issue. Remember that table from Chapter 6 showing the types of content offered in different journals? Table 8.2 shows that some of that content consists of review articles (shown underlined).

A few journals *only* publish review articles. The name of the journal usually indicates whether it has this specialization. For example, *Annual Review of Astronomy and Astrophysics* is a journal that comes out once annually and contains *only* review articles.

Table 8.2

New England Journal of Medicine	*Journal of Environmental Engineering*	*Geology*	*Nature Ecology and Evolution*
Perspective Original articles <u>Review articles</u> Editorials Correspondence	<u>State-of -the-art reviews</u> Technical articles Technical notes	Articles	Editorial Correspondence Comment & opinion News & views <u>Review articles</u> Research articles

To sum up, what should science and engineering students know about review articles?

1. Review articles are a type of journal article.
2. Review articles provide an overview of the current state of knowledge of a particular research area.
3. Review articles may be organized in the same way as a research article, or they may have their own unique organization.
4. There are different types of review articles that serve specific purposes. Some, like systematic reviews and meta-analyses, establish high levels of evidence that we're closing in on the right answer to a research question.
5. If you're trying to establish policies or best-practices, get guidance from systematic reviews and meta-analyses when you can.
6. Review articles are published in journals along with research articles; however, some journals *only* publish review articles.

What's Next?

We've spent a lot of time talking about how scientists and engineers communicate through research articles and review articles published in journals. But journals are not the only type of source for sharing research findings; **conference proceedings** offer another venue and we'll discuss them in the next chapter.

CHAPTER 9
CONFERENCE PROCEEDINGS AND CONFERENCE PAPERS

Conference proceedings, and the papers they contain, are another type of source heavily used by scientists and engineers. Like journals, these sources offer a venue for publishing research findings. This chapter will explain the characteristics conference papers have in common with journal articles, and how they're different. You'll also learn tips for finding conference papers and distinguishing them from other sources.

Conferences and the Publications That Emerge from Them

Before we examine proceedings and papers in detail, let's discuss the meetings—otherwise known as **conferences**—these sources emerge from. Most disciplines within science and engineering have biannual, annual, or biennial conferences at the regional, national, and sometimes the international level. These conferences might be sponsored by an educational institution, research institute, or government agency, but most often they're sponsored by a professional society.

A **professional society** is an organization committed to advancing a particular profession and supporting those working in that profession. Professionals can pay a fee to join these societies and then take advantage of the services the society offers. Typical services include providing news, running mailing lists, issuing publications, helping members organize into task forces, and coordinating meetings. Here are some examples of professional societies in the United States:

- American Chemical Society
- Geological Society of America
- American Psychological Association

And here are some professional societies that have an international scope:

- Institute of Electrical and Electronics Engineers (IEEE)
- Association for Computing Machinery (ACM)
- International Association for Plant Taxonomy

Gatherings of a professional society are usually called **meetings** or **conferences** and those two terms are often used interchangeably. However, professional gatherings are called by other names as well, depending on the number of attendees, the length of the meeting, and the reason for the meeting. For instance, a **seminar** is a small gathering where attendees focus on discussing a particular issue. A **workshop** is also a small gathering but with the objective of achieving a particular goal, usually with participants working together in groups. A **colloquium** usually has a series of experts or specialists giving talks. A **symposium** can be variable in size and may be focused on a particular topic. A **conference** can be quite large, and some have thousands of attendees. A **congress** is the largest type of meeting and may be international in scale.

The lines between these types of gatherings are blurry and their definitions are loose, so don't worry too much about distinguishing between them. To keep it simple, let's call all types of meetings "conferences" for the rest of this chapter. Here are a few examples of conferences that take place in the United States:

- Geological Society of America Annual Meeting
- American Society of Engineering Education Annual Conference
- Society of Vertebrate Paleontology Annual Meeting

Conferences provide researchers within a discipline the opportunity to meet and present their research, exchange ideas, and network for jobs and collaborators. Conferences may also provide events such as panel discussions or sessions that offer training. Many conferences also have a designated area for vendors where sales representatives can show attendees new products, technologies, and services. However, for most conferences, the main objective is to provide an organized space for attendees to present and share their current research.

Presenters will either deliver a talk on their research or show a poster depicting their research. Typically, a **talk** consists of a researcher (or team of researchers) showing a PowerPoint presentation about their research, with time set aside at the end for the audience to ask questions. A **poster** is a visual depiction of a research project. At certain times during the conference, presenters will stand near their poster and explain it to any conference attendees passing by who show an interest in it. This is called a **poster session**. By the way, the ability to put together a good poster is a fundamental skill for scientists and engineers, including students in science and engineering. Posters demonstrate your ability to condense your research into the essentials and explain it briefly, clearly, and with relevant visuals.

Not just anyone can present at a conference—it's restricted to attendees who submitted proposals for talks and posters to the conference organizers and had their proposals accepted. For some types of conferences, it may be fairly easy to have a proposal for a poster or talk accepted. For others, the acceptance rate may be very low; at these conferences it would be very prestigious to have your proposal for a talk or poster accepted.

To keep the conference and its attendees organized, a **program** is provided to attendees before the start of the conference. The program lists the talks that will take place, including the day, time, and location, as well as a short summary of what each talk will be about. Details about where and when posters can be viewed are also included in the program. The program allows attendees to plan their time and decide which talks and poster sessions to attend. Conference programs may be delivered to attendees in either print or electronic form (or both). For some conferences, a corresponding *publication* is produced. This publication can take a couple of different forms—for instance, it might be a volume of abstracts. **Abstracts** are summaries of talks given at a conference and are usually just a paragraph or two long.

A conference publication might also take the form of a **proceeding** instead of a volume of abstracts. For the rest of this chapter, that's what we'll discuss: conference proceedings and the papers within them.

An Overview of Conference Proceedings and Conference Papers

A **proceeding** contains the written-up versions of talks delivered at the conference, called **papers.** Table 9.1 shows how conference proceedings and their contents compare to journals and books.

Table 9.1

Proceedings contain	*Journals contain*	*Books contain*
papers*	**articles***	**chapters**

*Be aware that many people use the words "articles" and "papers" interchangeably.

Note that the name "proceeding" is not always consistently applied. Some conference organizers consider their program to be a proceeding. Also, if another type of publication emerges from a conference—such as a report on the achievements of the conference—it might be called a "proceeding." But usually when people refer to a "proceeding," they mean a publication that contains papers documenting research discussed at a conference.

A proceeding might be published by the organization, society, or agency sponsoring the conference. For example, the U.S. Department of Education sponsored this conference on wheeled mobility and also published the associated proceeding:

U.S. Department of Education. (2006). *Wheeled Mobility and Accessible Transportation, 2003 Conference Proceedings*. Washington, D.C.

Alternatively, the sponsoring organization might make an arrangement with a publishing company to print, distribute, and sell their proceeding.

By the way, a proceeding can be associated with any type of professional gathering, not just a conference. Below you see an example of a proceeding that came out of a workshop.

Sniezko, R. A., et al. (2012). *Proceedings of the Fourth International Workshop on the Genetics of Host-Parasite Interactions in Forestry: Disease and Insect Resistance in Forest Trees*. Gen. Tech. Rep. PSW-GTR-240. Forest Service, U.S. Department of Agriculture.

Publication and distribution of a proceeding might take place before the conference begins so that attendees can look over the papers and decide which talks they want to attend. However, it's more common for the proceeding to be published and distributed after the conference, as a formal record of what took place. The publication timing really depends on the norms of the society or organization that's meeting.

Let's move on to talk about the papers published within a conference proceeding. First, let's note some name variations. Papers within a proceeding are sometimes referred to as **conference papers**, **proceeding papers**, or **technical papers**, and you'll also hear them referred to as **articles**. The purpose of a conference paper is similar to the purpose of a research article published in a journal, and that is to describe discoveries, applications, or inventions associated with a research project. Typically, papers in a proceeding are written up much like a research article in a journal with a similar structure including an introduction, methods, results, and discussion (or variations on that structure).

A More In-Depth Comparison of Conference Papers and Journal Articles

While some conference papers are virtually identical in scope and layout to research articles, others have characteristics that set them apart. Consider that conferences offer presenters a chance to gather feedback on their research from other attendees. Naturally, feedback is particularly useful to have in the early stages of a research project. So, some of the talks at a conference may describe research that's still in a preliminary or exploratory stage, as this allows the presenter to gather feedback from attendees and use this feedback to further refine and shape their project. So, some conference papers describe early research activities and intentions, or preliminary results, rather than laying out a fully complete research project like you'd expect to see in a research article published in a journal. Conference papers may also include less formal content such as ideas for new methodologies, protocols, or best practices.

So conference papers (as well as conference posters) can offer a more flexible and informal venue for sharing research compared to a research article published in a journal. This can be especially useful for junior researchers, like students. Preparing a poster or talking about research ideas or early results at a conference can be simpler and more straightforward than publishing a journal article, plus it allows junior researchers to gather useful feedback.

Another difference between conference papers and journal articles is that conference papers tend to be shorter. Furthermore, the publication process for conference papers can be quicker. Whereas many conference papers undergo some sort of peer-review process prior to publication, it's usually on a faster schedule and with less back-and-forth to discuss revisions compared to journal articles. Also, there may be fewer peer reviewers looking at the paper. So, the peer-review process, if there is one, may be less stringent, and

this can reduce the amount of time it takes to publish a conference paper compared to a journal article. (If you are unfamiliar with the peer-review process, don't worry, you'll learn about it in Chapter 14.) For disciplines that advance rapidly, such as computer science, it's important to be able to publish rapidly. In fact, in computer science, conference proceedings are the primary means for disseminating research results, while journals are used much less frequently.

Of course, some conference papers are virtually identical to research articles in content, length, and rigor, plus they go might undergo a similar peer-review process. So, the exact nature of conference papers really depends on norms of the discipline and/or the conference. Over time you'll better understand the typical qualities and characteristics of conference papers in your discipline.

Finding Conference Papers

If you need to find a specific conference paper or if you're browsing for conference papers on a topic, the best place to start is usually your campus library's website. As you've learned in previous chapters, look for a search box on your library's homepage that invites you to search articles, books, media, and more, or has wording similar to that. This is called a **discovery tool** and you'll learn more about it in Chapter 20. You can enter a title search for a specific conference paper, or you can enter keywords to look for papers on a topic.

For example, suppose your electrical engineering professor told you to consult the paper below for a class project about collecting and storing accurate fingerprint data:

> Shehu, Y. I., et al. (2018). Detecting an alteration in biometric fingerprint databases. *Proceedings of the 2nd International Conference on Digital Signal Processing* (pp. 6–11).

Assuming your library offers a discovery tool search box right on their homepage, you can simply enter the title of the conference paper into the search box to see if your library has it. (Figure 3.1 shows how to search for a particular book; use this same strategy to search for a conference paper.) Once you've run your search, look for filters (which are usually to the left of your results) to limit your results to conference papers only, just like you learned to do for books in Figure 3.2.

If your library doesn't provide a discovery tool search box on their homepage, or if the search you conducted didn't find what you needed, then by now you should know to ask for help! Look for a link to a **chat service** on your library's homepage and initiate a chat to ask how to find conference papers. Or you can phone or email your library to ask—or visit in person.

But if you want to keep exploring on your own, and if you're looking for conference papers on an engineering-related topic, then consult Chapter 20 for tips on navigating your library's website to find **article databases** such as Compendex or **publisher databases** such as IEEE Xplore. Both have lots of conference papers in addition to articles.

You can also try searching Google or Google Scholar for a particular conference paper to see if you can find a free copy to read. Or you can go to the website of the body that organized the conference to see if they offer access to papers published within their conference proceedings. (Note that in addition to proceedings, they may also provide abstracts, posters, and even videos of talks from conferences.) But if you've exhausted all these options and still can't find what you need, then ask for help or consult the section below.

Dealing with Access Problems

If there's a conference paper you want to read, and you can't find it using library databases or by searching for a free copy on the internet, then request it through your **library's document delivery/interlibrary loan** service. You learned about this service in previous chapters, but as a reminder, it can supply you with sources that your library doesn't have. You should be able to find a link to this service on your campus library's homepage, but if you need help then ask library staff for assistance.

Distinguishing Conference Papers from Other Sources

The print version of a conference proceeding might look like a book, but you'll probably be able to tell it's a proceeding from its title. Proceeding titles usually contain the word "proceedings" and often include the conference name, date, and/or number, as you see in these examples:

- *Proceedings from the 17th European Conference on Soil Mechanics and Geotechnical Engineering*
- *WiSec 11 Proceedings of the Fourth ACM Conference on Wireless Network Security*
- *Proceedings of the Twentieth Annual Symposium on Sea Turtle Biology and Conservation, March 2000*

Distinguishing conference papers from other sources in an *online* environment can be harder, so let's go over some strategies for doing that. Suppose you're searching a library database and you aren't sure if it's presenting you with results that are conference papers. As you know by now, you can look for filters to refine your results to just conference papers, or you can check to see if the database identifies conference papers in your results. Figure 3.7 shows how some database results are labeled by source type, but look at Figure 9.1 for a fresh example of a result that turned up in a search on the topic of cybersecurity. Notice that the source type (Conference Paper) is provided at the bottom of the record.

Developing Cybersecurity Workforce: Introducing CyberSec Labs for Industry Standard Cybersecurity Training
Yasin Kose; Murat Ozer; Mehmet Bastug; Said Varlioglu; Oguzhan Basibuyuk; Hari Priya Ponnakanti
2021 International Conference on Computational Science and Computational Intelligence (CSCI)

Year: 2021 | Conference Paper | Publisher: IEEE

Figure 9.1

When conference papers are clearly labeled like this, you know when you're looking at one! But not all databases make it this easy. So, if you need to confirm you're looking at a conference paper, then try downloading it as a PDF, if you can. Then, look it over to see if there are clues in the header or footer indicating what publication it belongs to, just like you learned to with journal article PDFs in Chapter 6. If your PDF shows nothing in the header or footer, then look up the title in Google or Google Scholar to see what you can learn about its origins. Figure 9.2 shows you the process for looking for a particular title in Google Scholar to learn what type of source it is.

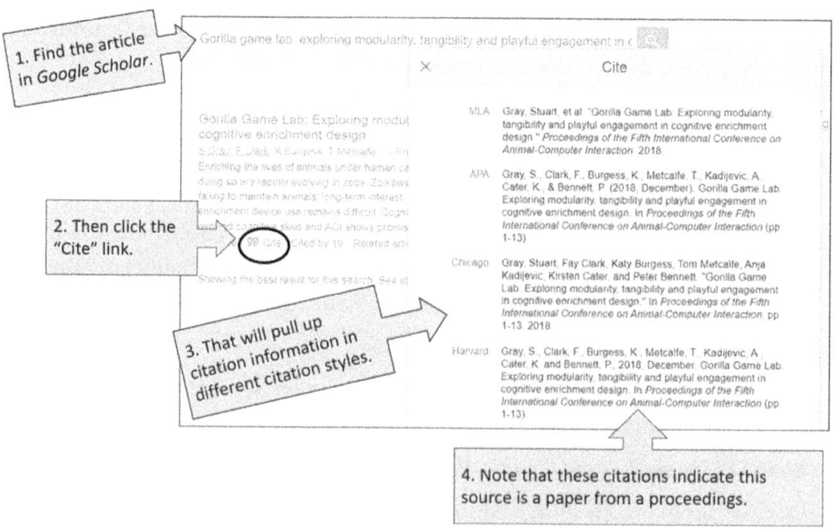

Figure 9.2

But here's a snag—sometimes conference papers are published in an issue of a journal rather than in a volume of proceedings. The journal issue is often called a **special issue** or **supplement issue** to show that it is different from other issues of that journal. An example is shown in Figure 9.3.

Figure 9.3

To sum up, conference papers are easily recognizable when they're published within a volume of proceedings, because proceedings have such distinctive titles. It would be hard to confuse a source published in *Proceedings of the Twentieth Annual Symposium on Sea Turtle Biology and Conservation, March 2000* as anything other than a conference paper. But just know that sometimes conference papers are published in journals and look just like journal articles; in those instances, just treat them as if they are journal articles.

Some Journals Have Titles that Sound like Conference Proceedings

You should also be aware there are a few journals that have titles that sound like conference proceedings. Here are a few examples:

- *Proceedings of the Royal Society of London*
- *Proceedings of the National Academy of Sciences (PNAS)*
- *Transactions of the American Entomological Society*

Although these titles *sound* like they're conference proceedings, they're actually journals. Most proceedings have a meeting number and/or date in the title, but these titles don't, which is a clue they're not conference proceedings. Why do these journal titles sound like proceedings? Because when they first started publication, they *were* proceedings. That is, the content in these journals originated from talks delivered at society conferences. Over time, though, they expanded to include content that was not associated with conferences, and so they morphed into journals while retaining their proceeding-like titles.

To sum up, what should science and engineering students know about conference papers?

1. Professional meetings (also called conferences) generate publications like abstracts, posters, or conference proceedings.
2. A conference proceeding is made up of papers that describe research presented in talks at the conference.
3. The research that conference papers describe is similar to what you find in research articles published in a journal. However, conference papers may be shorter, less formal, contain preliminary research results, etc.
4. Conference papers can be found in the internet, from library databases, as well as through library services such as document delivery/interlibrary loan. Libraries provide staff to help you locate conference papers when you need assistance.
5. Some conference proceedings are published in their own volume, while others might be published within an issue of a journal.

What's Next?

You've learned that journal articles and conference papers are two types of sources that allow researchers to share their research findings with others. But there are a couple more sources that present research findings; these sources are written by student researchers pursuing a post-graduate degree. We call these sources **dissertations** and **theses**, and you'll learn about them in the next chapter.

CHAPTER 10
DISSERTATIONS AND THESES

Dissertations and theses describe research conducted by graduate students. This chapter will describe the unique features of these sources and why they're useful to consult. You'll also learn how these sources can help launch your career as a researcher. Plus, you'll find out how to search for dissertations and theses and how to distinguish them from other sources.

An Overview of Dissertations and Theses

Dissertations and theses are written by graduate students. **Graduate students** already have an undergraduate degree and are now pursuing another course of study that will earn them an additional degree, such as a **Master of Science** degree (**MS** or **MSc**), a **Doctor of Medicine** degree (**MD**), or a **Doctor of Philosophy** degree (**PhD**)—also called a **doctorate** or **doctoral** degree. The exact title and corresponding initials of the degree depend on the course of study and can vary from country to country.

In the United States, doctoral students write up a **dissertation** as the culminating product of their research, whereas master's students write up a **thesis**. However, if you live outside the United States, a thesis might be the culminating work of a doctoral student, and the output of a master's student might be called a dissertation, exactly the reverse of how those terms are used in the United States.

Dissertations and theses describe research that a graduate student has been working on for several months or years. Figure 10.1 shows the typical content you'd find in a dissertation or thesis.

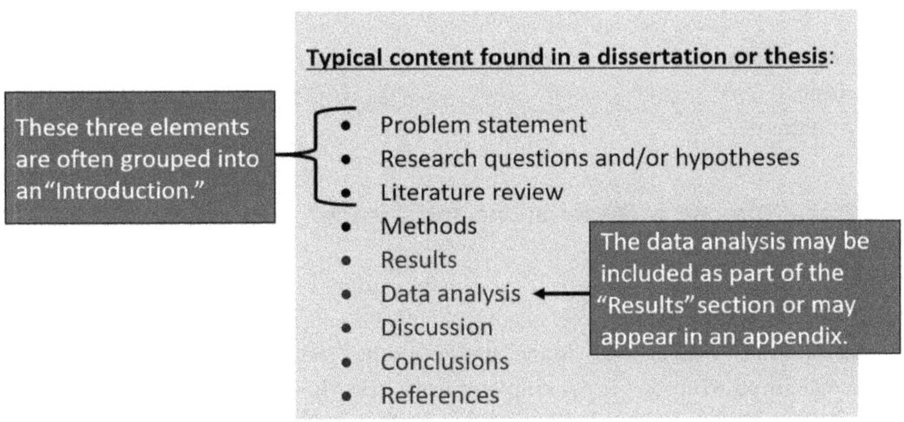

Figure 10.1

You might notice this content is similar to what you'd find in a research article published in a journal (refer to Chapter 7 for details). That's because dissertations and theses describe a research project and its results, just like research articles. However, dissertations and theses describe research projects with a broader scope. You can get a sense of this broader scope by glancing over the dissertation titles shown below.

Hardaker, P. (2018). *Using Artificial Intelligence to Improve the Control of Prosthetic Legs*. [Doctoral dissertation] De Montfort University.

Smith, A. L. (2017). *Contribution of Known and Novel DNA Repair Genes to Pancreatic Cancer Susceptibility*. [Doctoral dissertation] McGill University.

Coltrane, J. A. (2012). *Ecological and Physiological Adaptations of the Porcupine to Winter in Alaska*. [Doctoral dissertation] University of Alaska Fairbanks.

Valigura, J. (2019). *Seismic Assessment of Bridges with Hybrid Sliding-Rocking Columns*. [Doctoral dissertation] University of Colorado at Boulder.

Dissertations and theses are also much longer than research articles. That's partly because of their broader scope, but also because certain sections, like the literature review, methods, results, etc., are written up in more detail than what you'd typically find in a research article.

In fact, because dissertations and theses are so long and detailed, they tend to have correspondingly long lists of cited references at the end. So, if you're conducting research on a topic and happen upon a relevant thesis or dissertation, look at the cited references. The reference list might be a gold mine for finding additional sources on your topic.

Because dissertations tackle broader research projects (compared to research articles), they may feature research that's presented from an unusual perspective or that's unique in purview or framing. In other words, the broad scope of these sources allows authors to explore more dimensions of a research topic than would be possible in a research article. Below is an example of a dissertation that embodies some of these qualities:

Cashmore, D. D. (2020). *Completeness of the Dinosaur Fossil Record: Disentangling Geological and Anthropogenic Bias*. [Doctoral dissertation] University of Birmingham.

What's the Purpose of Dissertations and Theses?

These sources demonstrate a student's ability to conduct research, including compiling a literature review, designing and managing their own research project, and synthesizing their results and conclusions. Many students who are awarded a master's or PhD degree will go on to some sort of employment where they'll conduct research, so their dissertation or thesis shows they've gained the relevant skills to be researchers.

PhD students are especially likely to enter careers where they'll conduct research, and the research they conduct for their dissertation may help launch their career. How so? The first published research articles of a new researcher's career may be based on research they did in their dissertation. Many budding researchers will adapt and reshape content from their dissertation to turn it into one or more publishable research articles. For example, below is a research article that likely contains content and insights gained from this same student's dissertation that you saw in the previous section:

> Cashmore, D. D., et al. (2020). Ten more years of discovery: revisiting the quality of the sauropodomorph dinosaur fossil record. *Palaeontology*, *63*(6): 951–978.

Quality Issues

Because dissertations and theses represent *student* research, these sources may not be of the same caliber you'd get from a source that's authored by a seasoned expert or professional. So, if you're thinking about using information from a dissertation or thesis, first run some searches on the author to see if they've written and published any journal articles or conference papers on the same topic as their thesis or dissertation. If so, and assuming you can find the information you need in those sources, use and cite those sources instead. Journal articles have more authority and credibility than dissertations and theses because they've been put through a more rigorous peer-review and editing process; thus, it's preferable to cite them (when you can) rather than citing a thesis or dissertation.

Finding Dissertations and Theses

If you're looking for dissertations and theses, start your search on your campus library's website. Look for a search box on your library's homepage that invites you to search articles, books, media, and more, or has wording similar to that. This is called a **discovery tool** and you'll learn more about it in Chapter 20. You can enter a search for a specific dissertation/thesis or you can enter keywords to look for dissertations/theses on a topic. Once you've run your search, look for filters (which are usually to the left of your results) to limit your results to dissertations/theses only (see Figure 3.2 for an example of how these filters might look).

If you don't see a search box like this on your library's homepage, or if the search you conducted didn't find what you needed, then your best bet is to just ask for help. Look for a link to a **chat service** on your library's homepage and initiate a chat to ask how to find dissertations and theses. Or you can phone or email your library to ask, or visit in person.

You could also try consulting Chapter 20 for tips on navigating your library's website to find **content-specific databases** that specialize in providing dissertations and theses, such as ProQuest Dissertations & Theses Global. Plus you can also search for dissertations and theses using a variety of free databases available on the internet, including:

- NDLTD Global ETD Search
- Open Access Theses and Dissertations (OATD)
- PQDT OPEN
- EBSCO Open Dissertations
- DART-Europe E-theses Portal

You can easily find these databases with a Google search. A good place to start is probably NDLT Global ETD Search. By the way, "ETD" stands for Electronic Theses and Dissertations, in case you're wondering.

Dealing with Access Problems

If your search for a dissertation or thesis comes up short, you can try using your campus library's **document delivery/interlibrary loan** service to get what you need (consult previous chapters for more information about this service). However, this has a fairly low chance of success. Why? When a student completes their dissertation or thesis, only a few copies are printed—sometimes just two. The student gets a copy and a copy is provided to the library of the institution where the student completed their degree. Libraries don't like loaning their copies of dissertations and theses out to other libraries because they are the only institution with a copy and if it gets lost then *nobody* will have a copy. If you're having a hard time getting your hands on a dissertation or thesis you want to consult, then as a last resort you may be able to purchase your own copy from an online service such as ProQuest Dissertations Express.

Distinguishing Dissertations and Theses from Other Sources

Dissertations and theses are fairly easy to recognize in their print form. They're usually

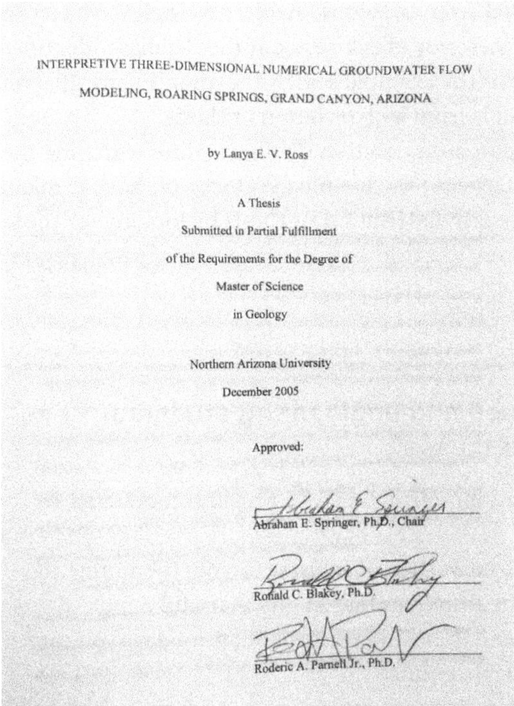

INTERPRETIVE THREE-DIMENSIONAL NUMERICAL GROUNDWATER FLOW MODELING, ROARING SPRINGS, GRAND CANYON, ARIZONA

by Lanya E. V. Ross

A Thesis

Submitted in Partial Fulfillment

of the Requirements for the Degree of

Master of Science

in Geology

Northern Arizona University

December 2005

Approved:

Abraham E. Springer, Ph.D., Chair

Ronald C. Blakey, Ph.D.

Roderic A. Parnell Jr., Ph.D.

Figure 10.2

printed on 8.5 × 11 paper and then bound in an unadorned, uniformly colored cover—often black.

Also, if you take a look inside a dissertation or thesis, the text is usually double-spaced. Plus, the title page will have certain unique elements such as the anticipated degree, the university where the student conducted the research, and the names of faculty members who reviewed and approved the dissertation or thesis. The title page shown in Figure 10.2 is a good example.

But distinguishing dissertations and theses from other sources in an *online* environment can be harder, so let's go over some strategies for doing that. Suppose you're searching a library database and you aren't sure if it's presenting you with results that

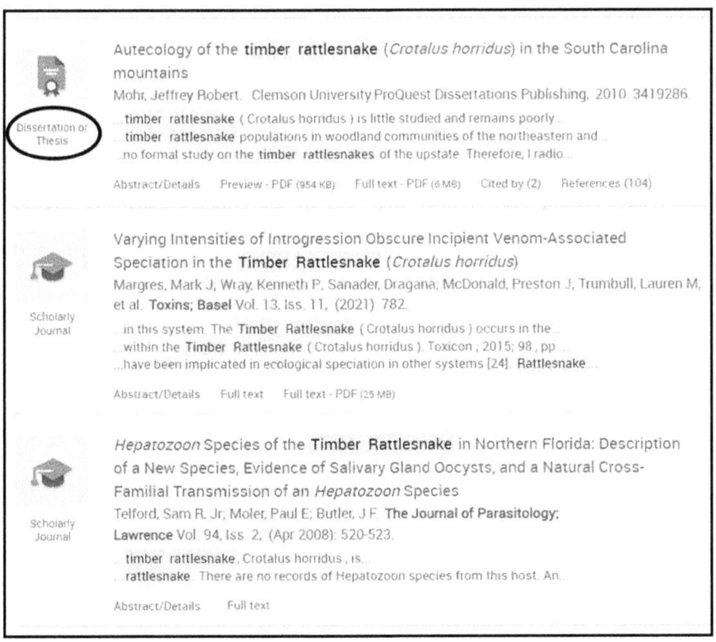

Figure 10.3

are dissertations or theses. You've learned from previous chapters to look for filters to refine your results to *just* dissertations/theses or check to see if the database identifies your results by source type, like you see in the example search shown in Figure 10.3 on timber rattlesnakes. Notice the first result is from a dissertation or thesis.

If, on the other hand, you happen upon a dissertation or thesis while searching the internet, then look for a PDF file you can view or download. Once you have it open, you should be able to confirm the source is a dissertation or thesis by looking for those unique, distinguishing elements it has on the title page, such as anticipated degree, university, etc.

To sum up, what should science and engineering students know about dissertations and theses?

1. Dissertations and theses describe long-term research projects conducted by graduate students.
2. If you find a dissertation or thesis on a topic of interest to you, then mine its cited references to discover additional sources that might be useful.
3. Only a few copies of dissertations and theses are printed, but nowadays you can often find them digitized and available online.
4. Dissertations and theses can be found on the internet and from library databases. Libraries provide staff to help you locate dissertations and theses when you need assistance.

What's Next?

In the next chapter we'll turn our attention to **websites** as a source of information. No doubt you're already quite familiar with navigating the internet and using websites, but understanding how the word "website" is used as a location and as a unique type of source might be new to you.

CHAPTER 11
WEBSITES

Websites have unique characteristics that distinguish them from the sources described in previous chapters. However, the word "website" is confusing because it's regularly used in two different ways—as a *location* and as a *type of source*. This chapter will explain that distinction. You'll also see examples of websites that are useful to scientists and engineers, and you'll learn to distinguish a website from other content located on the internet.

A Website Can Mean a Location

As you probably know, all content found on the internet has a location designated by an internet address called a **URL** (uniform resource locator). We commonly call this location a "website." For clarity, let's call it a **website-location**. So, all content on the internet is presented to you from a **website-location**. What types of content are we talking about? All of the following and more:

- online versions of books such as those found on Google Books
- online versions of newspapers like *The New York Times*
- online versions of magazines like *National Geographic*
- online versions of journals like *New England Journal of Medicine*
- the internet presence for a company, organization, educational institution, government agency, etc.
- blogs for fitness, gaming, cooking, auto repair, finance, news, etc.
- online reference tools like Wikipedia or Internet Movie Database

So, the content available from website-locations is highly variable and includes **traditional sources** like books, newspaper articles, magazine articles, and journal articles. But just because these traditional sources are available from website-locations doesn't make them **websites** (at least not in the way we'll define a "website" in the following section). Rather, they still maintain their integrity as traditional sources. In other words, a book is still a book, and a journal article is still a journal article, regardless of whether you're reading it in its print form or viewing it from a website-location.

Let's look at a specific example. Figure 11.1 shows an online journal article. This is the HTML version of the article, not the PDF version, but you could click the link

International Journal of Advanced Robotic Systems

Impact Factor: 2.3 / 5-Year Impact Factor: 2.3

Open access | Research article | First published online April 29, 2018

Development of a self-balancing robot with a control moment gyroscope

Ji-Hyun Park and Baek-Kyu Cho View all authors and affiliations

All Articles | https://doi.org/10.1177/1729881418770865

Contents | PDF / ePub | Cite article | Share options | Information, rights and permissions

Abstract

This study introduces a two-wheeled self-balancing mobile robot based on a control moment gyroscope module. Two-wheeled mobile robots are able to achieve better mobility and rotation in small spaces and to move faster than legged robots such as humanoid type robots. For this reason, the two-wheeled mobile robot is generally used as a mobile robot platform. However, to maintain its balance, the two-wheeled robot needs to use movements of its two wheels. When an unexpected disturbance affects the robot, the robot maintains its balance with movements of the wheels and tilting of the body. If the disturbance exceeds the response capability of the robot, the robot will lose its stability. At the same time, the safety of the robot may be put at risk by movements to maintain balance. To address these issues, a robot was designed with a control moment gyroscope module to improve balance while minimizing movement. When a disturbance is applied to the robot, the disturbance is estimated by a disturbance observer and the control moment gyroscope controller compensates the disturbance. Using the control moment gyroscope module, the robot can maintain balance with just small movements of its wheels. Improved performance and stability were verified with experiments and simulations.

Figure 11.1

just above the abstract to download the PDF. This journal article is available from this website-location: https://journals.sagepub.com/doi/full/10.1177/1729881418770865. However, that doesn't make it a website. Rather, it's a still a journal article.

A Website Can Mean a Type of Source

We also use the word "website" to refer to a particular *type of source* that's distinct from traditional sources like books, journal articles, etc. In this definition, a **website** is a collection of related, hyperlinked pages devoted to a specific topic or purpose, and created and maintained by an individual, organization, educational institution, company, etc. Note the word "maintained" in that definition. That word indicates that websites get updated; so, they're not static, they're dynamic—and that's one of their unique qualities.

Let's say you are looking at the website for the National Association of Geoscience Teachers (NAGT) available at https://nagt.org/index.html. This is a classic example of a website as a *type of source*. For clarity, let's call it a **website-source**. Website-sources are a recent addition to the array of information sources discussed in previous chapters. This type of source only came into existence with the emergence of the internet. Without the internet, this source couldn't exist; obviously you can't have dynamic, changing content and hyperlinks in a paper-based print source.

Even though the online journal article you saw in the previous section and the NAGT website mentioned above *both presented on website-locations*, only the NAGT website has the unique qualities of a *website-source*.

Now let's revisit that list of content you saw at the beginning of this chapter, and note in bold the content that qualifies as a website-source:

- online versions of books such as those found on Google
- online versions of newspapers like *The New York Times*
- online versions of magazines like *National Geographic*
- online versions of journals like *New England Journal of Medicine*
- **the internet presence for a company, organization, educational institution, government agency, etc.**
- **blogs for fitness, gaming, cooking, auto repair, finance, news, etc.**
- **online reference tools like Wikipedia or Internet Movie Database**

There are also website-sources for selling products, presenting reviews, providing entertainment, as well as providing individuals with a public place to share their hobbies, interests, rants, and other thoughts with the world. Plus, there are online forums, social networking sites, bookmarking sites, etc. So, the content available from website-sources is highly variable.

Quality, Utility, and Abundance of Website-Sources

The *quality* and *utility* of content available from website-sources is highly variable. Some website-sources contain incorrect, incomplete, oversimplified, or outdated information. Some are intentionally false and misleading—such as fake news sites. Other website-sources are harmless but inconsequential, such as someone's blog about their adventures with their dog. But other website-sources are extremely useful and beneficial. Here are some examples of useful website-sources that scientists and engineers might consult:

- Chemistry Now
- IUCN Red List of Threatened Species
- Mindat.org
- National Institute of Mental Health
- Stanford Engineering Everywhere
- WolframAlpha

Website-sources are also abundant. As you know, when you search Google for a topic you typically retrieve millions of results. Some of those results represent website-sources, whereas some represent traditional sources (like journal articles) delivered from website-locations—but regardless of the exact nature of the source, an internet search retrieves a *lot* of results.

Why is so much available on the internet? Because it's both cheap and easy for *anyone* to post content to the internet. To set up a website-source (like a blog) and post your thoughts, all you really need is an internet connection and access to a desktop computer, laptop, or handheld device. This makes it possible for millions of people to present content on the web, and most of that content isn't ever going to be useful to you.

Domain Names Help Keep Websites Organized ... Sort of, but Not Really

All content presented on the internet is assigned a domain name like **.com**, **.org**, **.edu**, **.gov**, **.int**, etc., and these domain names show up in the website's URL. Domain names help delineate internet content into broad categories, but not in a way that's particularly organized, consistent, or useful for promoting information literacy.

The purpose of domain names is to signify the *type of entity* that's putting forward content on the internet:

- **Companies (.com)** such as Tesla or Elsevier Publishing
- **Organizations (.org)** such as the National Association of Geoscience Teachers (NAGT)
- **U.S. educational institutions (.edu)** such as Harvard University
- **U.S. government agencies (.gov)** such as the National Park Service
- **Intergovernmental organizations (.int)** such as the World Health Organization

However, for a lot of information on the internet, knowing whether it's coming from a **.com**, **.edu**, or **.org** domain isn't really that useful. Rather, it would be more useful to know something about the *quality* or *function* of the content, such as: is it a social networking site, or a scholarly journal, or a database? If internet content was delineated into narrower and more descriptive domain-name categories, and if those categories captured qualities and/or functions, then it would be much easier for everyone to understand what they're viewing on the internet and how to use that content.

For example, consider the website for the American Sociological Association (www.asanet.org), which presents information about this organization. Then consider the website for HathiTrust (www.hathitrust.org), which is primarily used as a database for finding digitized books. Both sites have a **.org** domain. But can you see how it would be easier to understand the functions of these two resources if the HathiTrust website had been given the domain **.dbs** to signify its function as a database?

Or, what if you came across the website for the scholarly, peer-reviewed journal *Cell Genomics* and rather than having a **.com** domain (which is what it has) it had a **.jnl** domain to signify that it's a journal. Instead, this scholarly journal shares the same domain as Facebook, Amazon, HuffPost, Wikipedia, YouTube, Walmart, Reddit, etc.

Assigning more specific, informative domain names would also make it easier to properly **cite** the content you're viewing. If you *know* you're looking at a journal article because it has a **.jnl** domain name, then you'd know to cite it as a journal article instead of as a website. Unfortunately, it would take a lot of effort to reorganize, define, and assign new and improved domain names, which means we're probably stuck with things the way they are.

Another problem with domain names is that they don't even do a great job of capturing different types of entities. For instance, the **.org** domain used to be restricted to internet content coming from non-profit organizations, but this restriction was removed in 2019 and now any entity can register for this domain. And to make things even more confusing, many websites outside the United States use two-letter **country code domain names** to signify what country the website is coming from. However, the rules dictating when a website should get a country code domain name (instead of some other domain name) are different in each country.

Finding Websites

The most common way to search, find, and access internet content is by using an internet search engine like Google. You can search for a known site, like the website for Wikipedia, or you can put in keywords to search for a topic. When you search for a topic—such as **coral reef conservation**—your results will likely offer links to a combination of sources, some will be websites and some will be traditional sources delivered from a website location, like you see in Figure 11.2.

Internet search engines (like Google) work well for finding information on the "surface web." The **surface web** consists of free website-sources like personal blogs,

Figure 11.2

company websites, organization websites, Wikipedia entries, etc., as well as content like online magazine articles and online journal articles. However, sometimes an internet search will lead you to a page that *represents* content, but doesn't allow you to view that content without paying, just like you see in this **paywall** page which invites you to pay to view a journal article about porcupines in *Journal of Zoology* (Figure 11.3).

If you reach a paywall like this and you don't want to pay, then check to see if the source is available at your library, just like you learned to do in previous chapters. And don't forget to ask library staff for help if you run into problems!

Just so you know, internet search engines like Google can't find everything on the internet. For example, content that requires logging in or is restricted by privacy settings may not show up in your Google results. Content like this makes up the **deep web**—in contrast to the **surface web**. For example, if you use a learning management system at your campus—such as Blackboard, Canvas, or Brightspace—then content in those systems is part of the deep web and not retrievable in a Google search.

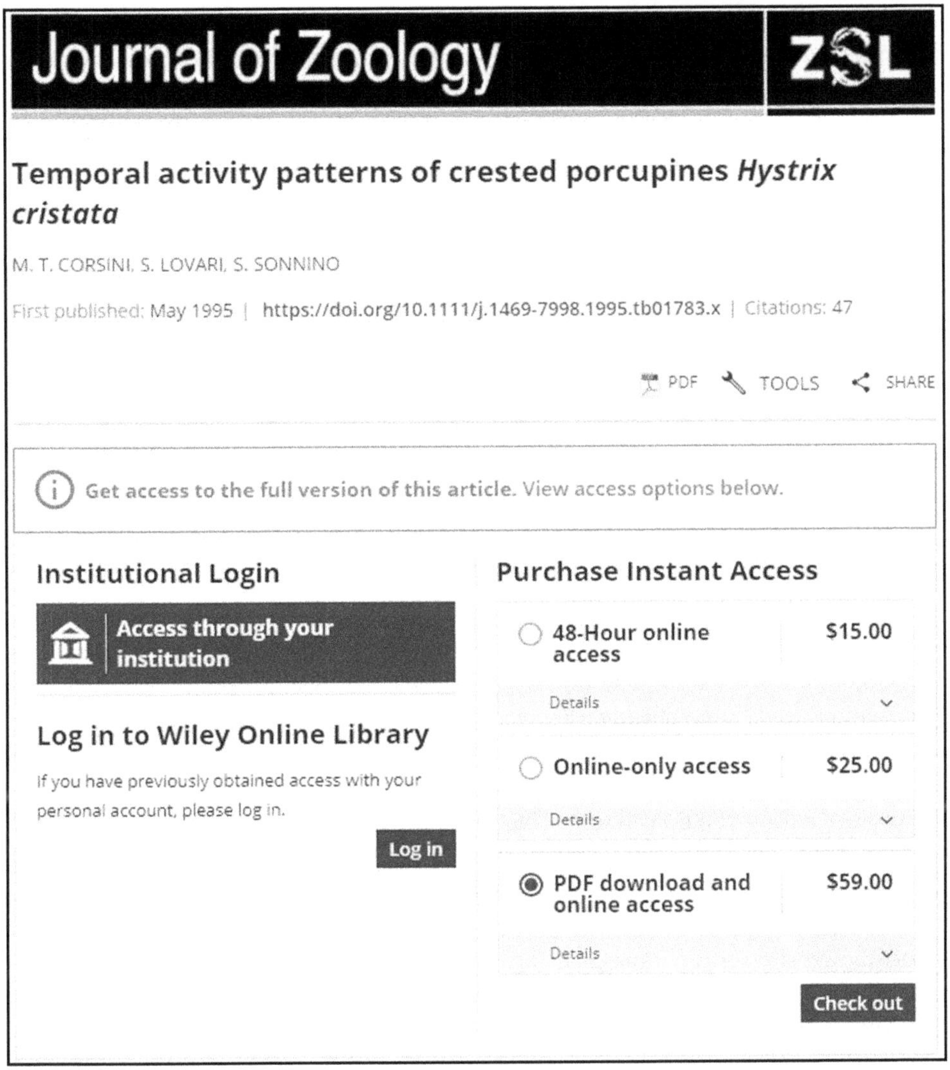

Figure 11.3

Are You in the Habit of Using the Internet for All Your Information Needs?

If you conduct internet searches for all your information needs, and if you give up on your search when you can't immediately find what you need, then try to break this habit! Lots of excellent information isn't digitized, or is protected by copyright, or requires payment, etc., and thus isn't freely available on the internet. There may be a perfect source out there to resolve your information need, but to find it you may need to look elsewhere than the internet, and the best place to start is usually a library. Libraries purchase or subscribe to all sorts of high-quality information sources that are either not available on the internet at all, or not *freely* available on the internet.

Distinguishing Website-Sources from Other Sources

As you saw earlier, when you search Google for information on a topic, you'll likely get a mix of website-sources and traditional sources in your results, and it'll take practice to distinguish them. Take another look at Figure 11.2, which depicts results from a Google search on **coral reef conservation**. Note that the first result leads to a website-source.

How would you know that result leads to a website-source? The name of the result (Coral Reef Alliance), as well as its URL and the information provided in the description, all indicate the result will lead to the homepage for an organization called the Coral Reef Alliance. Also, if you were to click on that result and examine it more closely, you'd notice it doesn't have the navigational features or content typical for online books, journal articles, newspaper articles, or magazine articles.

Does it matter whether you're able to distinguish website-sources from other types of sources you find online like books and journal articles? It does! You can resolve information needs more quickly when you know which type of source to match to your particular need. However, this requires the ability to distinguish traditional sources from each other as well as from website-sources.

Another reason you'll need to be able to distinguish different source types is to **cite** them properly when you use them in your writing. We'll talk more about citing in Chapters 29–31, but for now just note there's a specific format you must follow to create a citation and that format varies depending on whether you're citing a journal article, a book, a website-source, a magazine article, etc. Lots of students make the mistake of thinking that anything they find on the internet should be cited like a website-source. Not true—see for yourself:

Here's the typical format for citing a website-source:

> Author's Last name, First name. "Title of the Webpage." Title of the Website, date of publication, URL, date accessed.

And here's the typical format for citing a journal article coming from a website-location:

> Author's Last name, First name. (Date) "Title of Article." *Journal Title*, vol. #, issue #, page number(s), DOI (if available).

To sum up, what should science and engineering students know about website-sources?

1. We use the word "website" in two ways: as a location and as a specific type of source.
2. A website-source is a collection of related, linked pages devoted to a specific topic or purpose, and created and maintained by an individual, organization, educational institution, company, etc.
3. Website-sources are highly variable, abundant, and range widely in both quality and utility.
4. Lots of excellent information is not freely available on the internet, so you may not want to rely on the internet for all your information needs.
5. To properly cite sources, you need to be able to distinguish the sources you find online from each other.

What's Next?

In the next chapter we will examine **data**. Data have become a more heavily used source ever since policies were put in place to better manage data storage and sharing.

CHAPTER 12
DATA

Scientists and engineers create and collect data, but they also use data as a source—and it's never been easier to do so. This chapter will describe how the use and reuse of data has changed over time. You'll also learn about the obligations scientists and engineers have to organize, store, and share the data they've gathered.

An Overview of Data

Data are units of information collected using instruments or direct observation. Data are often numeric (such as measurements), but data can also be words that name or describe something, as well as pictures, sounds, etc. Research institutes, government agencies, and various other entities collect data. For example, the United States National Weather Service collects and records weather data, while the United States Census Bureau collects and records demographic data.

Individual researchers or research teams also collect data. For example, the researchers who wrote the article below gathered data on different ant species observed on islands in the Boston Harbor:

> Clark, A. T., et al. (2011). The effects of biogeography on ant diversity and activity on the Boston Harbor Islands, Massachusetts, USA. *PLoS One*, *6*(11): e28045.

In another example, the article below describes research on the walking gaits of lower-limb amputees. As amputees walked using their prostheses, the authors of this article collected data on the transfer of energy between legs. The data they gathered will inform the development of improved prosthetic legs that will allow amputees to walk with a more natural and efficient gait.

> Weinert-Aplin, R. A., et al. (2017). Energy flow analysis of amputee walking shows a proximally-directed transfer of energy in intact limbs, compared to a distally-directed transfer in prosthetic limbs at push-off. *Medical Engineering and Physics*, *39*: 73–82.

In the past, data collected from studies like these wasn't always saved and stored in a way that could be easily accessed and reused by other researchers. However, over the last twenty years or so, space to store digitized data has become cheaper and more readily available. At the same time, the internet has expanded our ability to share data. Together these two trends have resulted in an increased interest in ensuring that data are collected, organized, stored, and shared in ways that are optimally accessible and reusable. The expansion in access to data means that researchers can now find and use collections of data, also known as **data sets,** in all sorts of new and different ways, thus expanding our capacity for discovery. Everyone benefits when data are readily available for reuse and re-analysis. For example, take a look at the webpage shown in Figure 12.1 from the University of California Irvine's (UCI) Machine Learning Repository. This page offers a link to download data on concrete compressive strength.

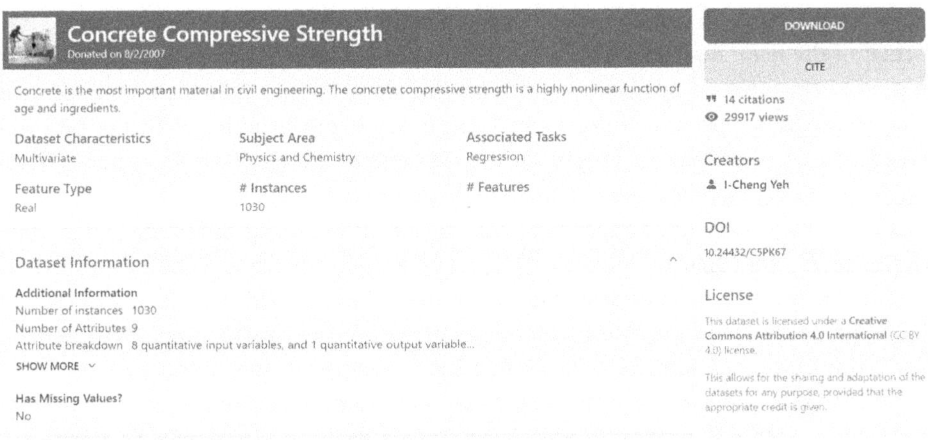

Figure 12.1

This data set, provided by I-Cheng Yeh at Chung-Hua University, has been used and cited in other studies, such as these two journal articles:

Nguyen, H., et al. (2021). Efficient machine learning models for prediction of concrete strengths. *Construction and Building Materials, 266*: 120950.

Vakharia, V. & Gujar, R. (2019). Prediction of compressive strength and Portland cement composition using cross-validation and feature ranking techniques. *Construction and Building Materials, 225*: 292–301.

The increased access to data didn't just happen spontaneously. Rather, it required the implementation of data storage and sharing policies. In the United States, government agencies played a key role in this process. How did it come about? The story starts with the way researchers are funded. In the United States, researchers can submit research funding proposals to certain government agencies such as the Department of Energy, the National Institutes of Health, and the National Science Foundation. If the agency decides to fund the research, the researcher will use the funds to cover costs associated with their research project, such as buying instruments, equipment, and supplies, as well as hiring staff and traveling to research sites, etc. In return, researchers must fulfill certain requirements established by the funding agency, such as reporting back on their research progress, publishing the results of their research, etc. A more recent requirement is that researchers must make the data generated from their research available to other researchers. Funding agencies have established **mandates** to enforce data sharing, and researchers who don't comply with these mandates will not get funded or may have their funding withdrawn.

So nowadays when researchers submit an application to a government agency for research funding, they must include a detailed plan for how they'll store and share the data generated by their research project. This is called a **Data Management Plan (DMP)** and it must include details about the type of data the research project will generate, the format of the data, how data will be labeled and organized, as well as any other details that will help others easily access and reuse the data. Within their data management plan, researchers must also address any data privacy issues, as well as how the data will be stored and preserved for long-term use. Data were a huge mess and often not re-usable before data management plans were implemented.

It's not just government agencies who are promoting and enforcing better data management and sharing practices. An increasing number of non-governmental funding agencies, as well as academic institutions, have joined the cause and created their own data storage and sharing mandates to further increase the availability of data. Nor is this movement limited to the United States—rather, it's global in scope.

Journal publishers are also helping. Many now require authors to provide a **Data Availability Statement** letting readers know where data associated with an article can be accessed. Furthermore, some journal publishers provide links to relevant data sets right on an article's webpage. Look for links to **Supplementary Material** or **Supporting Information** (or similarly worded links) to find those data sets. For example, the authors of the journal article below collected data on tooth marks found on Neolithic bone spoons:

> Stefanović, S., et al. (2019). Bone spoons for prehistoric babies: detection of human teeth marks on the Neolithic artefacts from the site Grad-Starčevo (Serbia). *PLoS One, 14*(12): e0225713.

If you went to the webpage where that article is located (Figure 12.2), toward the bottom you'd notice a section for "Supporting Information" with links to several data sets.

In the last several years, brand-new journals have started up that focus on providing articles about data. These articles describe and disseminate data as well as tackle issues relevant to the sharing and use of data. Here are the titles of some of those journals:

- *Data*
- *Data in Brief*
- *Data Science Journal*
- *Scientific Data*

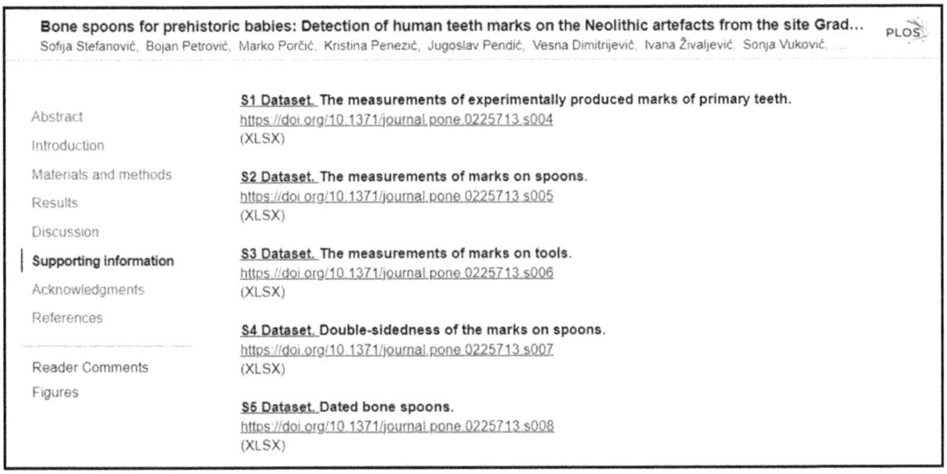
Bone spoons for prehistoric babies: Detection of human teeth marks on the Neolithic artefacts from the site Grad... PLOS

Sofija Stefanović, Bojan Petrović, Marko Porčić, Kristina Penezić, Jugoslav Pendić, Vesna Dimitrijević, Ivana Živaljević, Sonja Vuković, ...

Abstract
Introduction
Materials and methods
Results
Discussion
Supporting information
Acknowledgments
References

Reader Comments
Figures

S1 Dataset. The measurements of experimentally produced marks of primary teeth.
https://doi.org/10.1371/journal.pone.0225713.s004
(XLSX)

S2 Dataset. The measurements of marks on spoons.
https://doi.org/10.1371/journal.pone.0225713.s005
(XLSX)

S3 Dataset. The measurements of marks on tools.
https://doi.org/10.1371/journal.pone.0225713.s006
(XLSX)

S4 Dataset. Double-sidedness of the marks on spoons.
https://doi.org/10.1371/journal.pone.0225713.s007
(XLSX)

S5 Dataset. Dated bone spoons.
https://doi.org/10.1371/journal.pone.0225713.s008
(XLSX)

Figure 12.2

The emergence of these new journals shows how important data—as a source—has become.

Data Sharing Promotes Transparency

Sharing data from a research study has another important benefit; it makes the study more transparent and replicable. If there's a question about the validity of the study, others can reproduce it to see if they get the same data and arrive at the same conclusions. So, making data from a study available to others allows others to more easily verify the study's findings.

Be Prepared to Properly Manage Your Data Right from the Start

Once you're immersed in your career and embarking on research projects, make sure you know about any data mandates you must comply with, especially if you intend to publish the results of your research. If your research project is funded by government agencies or private foundations, then learn about the funder's data management policies. You should also be aware of your employer's data management policies and the data policies of journals where you might publish your results.

If you practice good data collection and documentation habits right from the start of your project, then it'll be easy to comply with data mandates. However, if you wait until the end of your project to organize your data and prepare them for sharing, you may find you've forgotten important details about the data. This could make it difficult to comply with data management and sharing mandates.

Data Sharing and Privacy Issues

It's common to hear about breaches of data privacy in the news, and the newspaper article below offers a good example:

> Hauser, C. (2020). EasyJet says cyberattack stole data of 9 million customers. *The New York Times.* https://www.nytimes.com/2020/05/19/business/easyjet-hacked.html

Some researchers, such as medical and social scientists, regularly conduct studies on human subjects, and that means collecting personal data on those subjects. So, you may be wondering how those researchers protect those data to keep them private. Aren't their data sets likely to be hacked and used for criminal purposes such as identify theft? Aren't the chances of data breaches even higher when researchers share their data with other researchers?

The data researchers collect on human subjects usually doesn't include social security numbers or credit card numbers that could be used for identity theft, but the data must still be protected. In fact, there are measures in place to keep personal data private in *all* stages of research, including as the data are collected, while they are stored, and any time the data are shared.

So, what are some of these measures? For some research studies, personal data that are collected are anonymous and can't be connected to the individual who provided it. In cases where human subjects' names *are* collected along with other personal data, the names must be coded (usually with numbers or some alphanumeric system). Any direct or indirect codes that could be linked to particular individuals must be de-identified or destroyed before the data can be shared. If data sharing would violate privacy rules established by an Institutional Review Board (explained below), or violate any local, state, or federal laws, then the researcher will be excused from complying with data-sharing mandates. Researchers must also store data about human subjects on secure servers that are password protected, and any data that are especially sensitive must be encrypted.

In the United States, regulations for protecting the privacy of human research data were established in 1974 with the passing of the National Research Act. This act led to the formation of Institutional Review Boards at all institutions or organizations in the United States that receive federal research funding. **Institutional Review Boards (IRBs)** are committees that review research proposals to ensure any research involving humans or other animals is conducted in an ethical manner. Institutional Review Boards can reject research proposals that do not meet appropriate ethical standards. Because privacy is an ethical issue for human subjects, these boards ensure that research projects don't go forward unless there are adequate measures in place to protect and secure personal data. As for *sharing* data on human subjects, there are guidelines for that as well. Those guidelines can be found on the websites of federal funding agencies such as the National Institutes of Health and the National Science Foundation.

Finding Data

If you're looking for data sets on a topic, there are several places you can search. You can start on your library's home page—just look for a search box that invites you to search articles, books, media, and more, or has wording similar to that. This is called a **discovery tool** and you'll learn more about it in Chapter 20. Enter keywords representing the subject matter of the data you'd like to find, then when you get your results, filter them to data sets, if that option is available. You saw how to filter results to books only in Figure 3.2—just use the same strategy to filter to data sets.

If your library doesn't provide a discovery tool search box on their homepage, or if the search you conducted didn't find what you needed, then there are other options for finding data. For example, you could consult Chapter 20 for tips on navigating your library's website to find **content-specific databases** that specialize in providing data such as Data Citation Index, Data Planet Statistical Datasets, and ICPSR. Or, you could try finding data sets by searching any or all of the free databases shown below, all of which are easy to find with a Google search:

- DATA.gov (for data collected by U.S. government agencies)
- Google Dataset Search
- Harvard Dataverse
- Re3data.org

There may also be instances where you come across a research article or proceeding paper that describes data of interest to you. If so, look it over to see if there is a data availability statement. For example, Figure 12.3 shows an excerpt of information taken from the first page of this journal article:

Titus, M., et al. (2021). Unsupervised manifold learning of collective behavior. *PLoS Computational Biology*, *17*(2): e1007811.

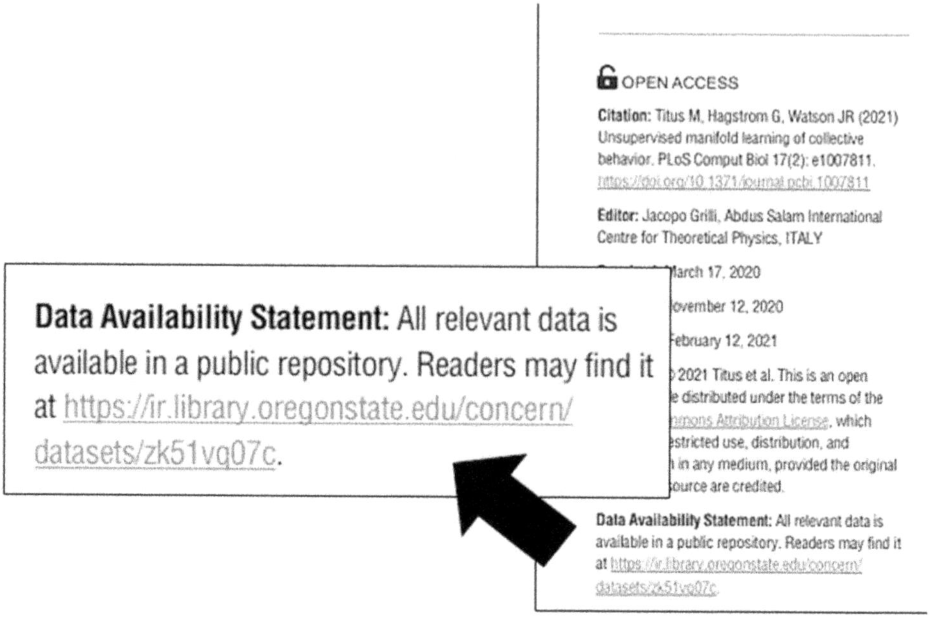

Figure 12.3

If you can't find a data availability statement, then try looking up the title of the article in Google or Google Scholar to see if you can find the HTML version of the article on the publisher's website. If you find it, scroll through the webpage to see if there are links that lead you to data sets (see Figure 6.5). Again, these links will likely be found under headings like **Supplementary Material** or **Supporting Information**—just like you saw in the article about Neolithic bone spoons provided earlier in this chapter.

If you still can't find the data you need, then try emailing the authors to ask if they can provide you with their raw data or direct you to where their raw data reside. Author's emails usually appear somewhere on the first page of the article, often right under the author's names.

Dealing with Access Problems

If there's a data set you need, and your library can't provide it, then your best bet will be to track down the author/creator of the data to ask if they can provide it. (Most of the time data sets can't be obtained through your library's **document delivery/interlibrary loan** service.)

Citing Data and Getting Permission to Use It

Before you use other researchers' data, you may need to seek formal permission from them. Furthermore, your local Institutional Review Board may require you to present a letter of permission for any data set you intend to use. Assuming you get permission to use a data set, you'll need to acknowledge your use of that data set in a citation when you write up your research project. In other words, when you use others' data you must cite them, just like you would any other source. You'll learn more about citing and citations in Chapters 29–31.

Distinguishing Data from Other Sources

Data are pretty easy to distinguish from other sources because they consist of discrete bits of information rather than paragraphs of written text. Also, data are usually stored in files that have their own unique **filename extensions**, also called **file extensions**. A file extension is the suffix at the end of a file name that identifies the type of file it is. For example, a Microsoft Word document has a ".doc" extension and PDF files have a ".pdf" extension. Data files usually have extensions such as .csv, .xml, .dat, .log, .sql, .xlsx, etc. And particular types of data, such as geographic data, have their own array of file extensions such as .gpx, .shp, .dbf, etc.

How Do *Students* Engage with Data?

As a student, you may not be ready to conduct a research project, publish your findings, and share the data generated by your research project. But you can still find and use data, plus you can help generate research data! If you're an undergraduate student who is interested in helping generate research data, seek opportunities to work with faculty or graduate students who need assistance collecting data for their research projects. You can also explore opportunities to collect and/or generate data through citizen science projects. Citizen science is a movement to crowdsource the collection and/or analysis of data by members of the general public—including students. Research projects involving citizen scientists are usually run by expert scientists and may even lead to published research, like in this article:

> Zipf, L., et al. (2020). Citizen scientists and university students monitor noise pollution in cities and protected areas with smartphones. *PLoS One*, *15*(9): e0236785.

So, even if you're not yet ready to publish and share data, you can still help generate it!

To sum up, what should science and engineering students know about data?

1. Data have become increasingly accessible and reusable over time.
2. In your career, you may need to comply with mandates for managing and sharing your data.
3. If you collect data on human subjects, you'll need to ensure the privacy of those data.
4. Data associated with a particular journal article may be available on the journal article's website.
5. You can find data sets using free internet databases as well as special databases that campus libraries provide.
6. You may need to seek permission to use data sets, and you should cite any data sets you use.

What's Next?

You're nearly done with Part II of this book—you just have one more chapter! The final chapter is about **patents**, **standards**, and **codes**, so it'll be especially relevant to engineers.

CHAPTER 13
PATENTS, STANDARDS, AND CODES

Patents, standards, and codes are particularly important sources for engineers, so if that's your discipline, this is an important chapter! You'll learn what these sources are, how they're used, and why they're important. You'll also learn how to find and access them.

An Overview of Patents

Suppose you came up with an idea for an invention and then put in long hours to design it, plus you used your own money to build and test prototypes. You'd have little incentive to exert this effort unless you *owned* your invention and could market and sell it for profit—thus paying yourself back for your creativity, time, and expenses. That's why we have intellectual property laws in place. **Intellectual property** refers to any sort of intellectual or creative output that's protected by law from unauthorized use by others.

There are three types of intellectual property safeguarded by three corresponding types of protections. Original works, such as books, songs, or photos are protected by **copyrights**. Phrases, logos, and designs that identify a company or product are protected by **trademarks**. And new products and inventions, such as toys, machines, pharmaceutical drugs, etc., are protected by **patents**. Patents record the details of an invention along with the person or company who invented it. As such, they function as a certificate of ownership for the inventor, providing formal documentation that the invention is their property.

In the United States, patents are issued by the United States Patent and Trademark Office (USPTO). But what rights does a patent issued by the USPTO give the owner? According to the USPTO, a patent provides the following:

> ... the right to exclude others from making, using, offering for sale, or selling the invention in the United States or importing the invention into the United States. What is granted is not the right to make, use, offer for sale, sell or import, but the right to exclude others from making, using, offering for sale, selling or importing the invention.

Patents don't last forever—rather, they are only good for fifteen to twenty years from the date the patent was filed (the exact amount of time depends on the type of patent), and only if the inventor has paid the required fees for registering and maintaining the patent. However, there are circumstances where the length of patent protection can be extended beyond twenty years.

In Figure 13.1 you see an example of a patent. The inventors called this patent "Sampling probe, gripper, and interface for laboratory sample management system." This is a utility patent, one of three types of patents. **Utility patents** are issued for *objects* such as machines, substances, or gadgets, as well as for *processes* that accomplish something.

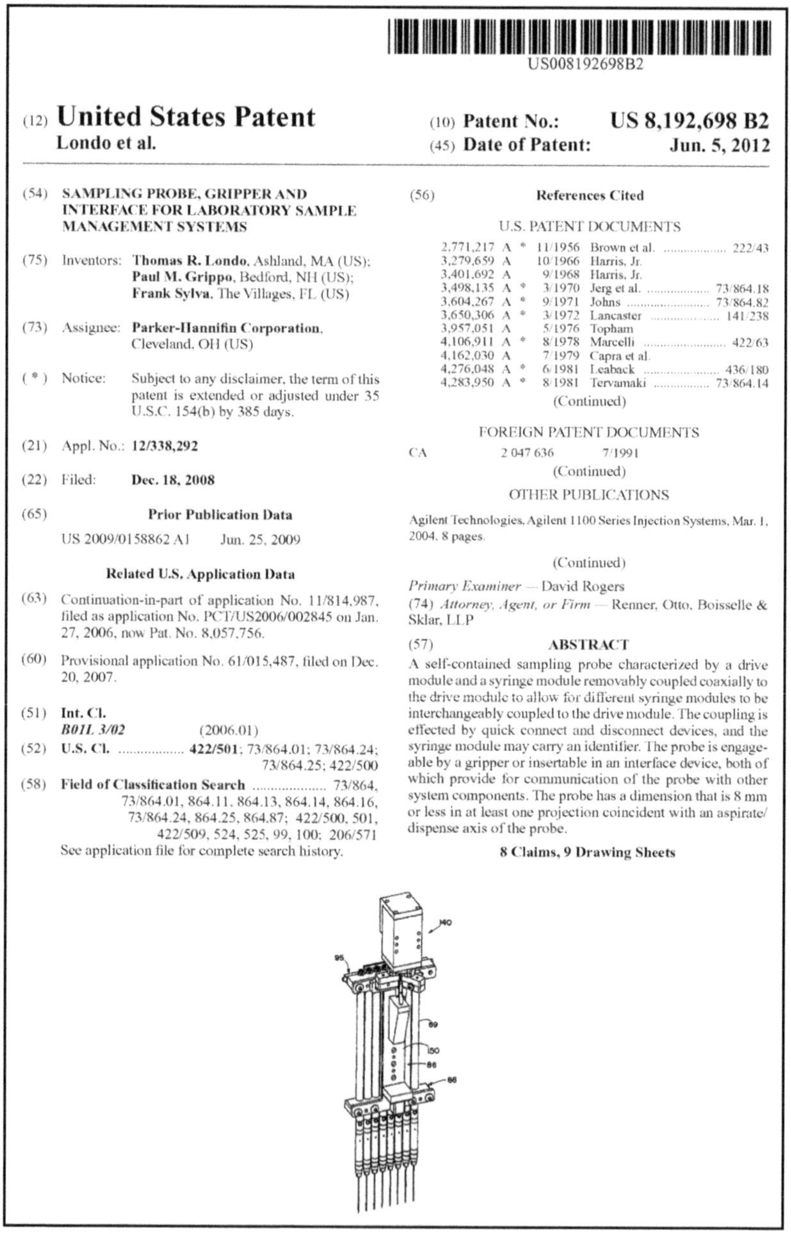

US008192698B2

(12) **United States Patent**
Londo et al.

(10) **Patent No.:** **US 8,192,698 B2**
(45) **Date of Patent:** **Jun. 5, 2012**

(54) **SAMPLING PROBE, GRIPPER AND INTERFACE FOR LABORATORY SAMPLE MANAGEMENT SYSTEMS**

(75) Inventors: **Thomas R. Londo**, Ashland, MA (US); **Paul M. Grippo**, Bedford, NH (US); **Frank Sylva**, The Villages, FL (US)

(73) Assignee: **Parker-Hannifin Corporation**, Cleveland, OH (US)

(*) Notice: Subject to any disclaimer, the term of this patent is extended or adjusted under 35 U.S.C. 154(b) by 385 days.

(21) Appl. No.: **12/338,292**

(22) Filed: **Dec. 18, 2008**

(65) **Prior Publication Data**

US 2009/0158862 A1 Jun. 25, 2009

Related U.S. Application Data

(63) Continuation-in-part of application No. 11/814,987, filed as application No. PCT/US2006/002845 on Jan. 27, 2006, now Pat. No. 8,057,756.

(60) Provisional application No. 61/015,487, filed on Dec. 20, 2007.

(51) **Int. Cl.**
B01L 3/02 (2006.01)

(52) **U.S. Cl.** **422/501**; 73/864.01; 73/864.24; 73/864.25; 422/500

(58) **Field of Classification Search** 73/864, 73/864.01, 864.11, 864.13, 864.14, 864.16, 73/864.24, 864.25, 864.87; 422/500, 501, 422/509, 524, 525, 99, 100; 206/571
See application file for complete search history.

(56) **References Cited**

U.S. PATENT DOCUMENTS

2,771,217	A	*	11/1956	Brown et al.	222/43
3,279,659	A		10/1966	Harris, Jr.	
3,401,692	A		9/1968	Harris, Jr.	
3,498,135	A	*	3/1970	Jerg et al.	73/864.18
3,604,267	A	*	9/1971	Johns	73/864.82
3,650,306	A	*	3/1972	Lancaster	141/238
3,957,051	A		5/1976	Topham	
4,106,911	A	*	8/1978	Marcelli	422/63
4,162,030	A		7/1979	Capra et al.	
4,276,048	A	*	6/1981	Leaback	436/180
4,283,950	A	*	8/1981	Tervamaki	73/864.14

(Continued)

FOREIGN PATENT DOCUMENTS

CA 2 047 636 7/1991

(Continued)

OTHER PUBLICATIONS

Agilent Technologies, Agilent 1100 Series Injection Systems, Mar. 1, 2004, 8 pages.

(Continued)

Primary Examiner — David Rogers
(74) *Attorney, Agent, or Firm* — Renner, Otto, Boisselle & Sklar, LLP

(57) **ABSTRACT**

A self-contained sampling probe characterized by a drive module and a syringe module removably coupled coaxially to the drive module to allow for different syringe modules to be interchangeably coupled to the drive module. The coupling is effected by quick connect and disconnect devices, and the syringe module may carry an identifier. The probe is engageable by a gripper or insertable in an interface device, both of which provide for communication of the probe with other system components. The probe has a dimension that is 8 mm or less in at least one projection coincident with an aspirate/dispense axis of the probe.

8 Claims, 9 Drawing Sheets

Figure 13.1

Utility patents can also be issued for *improvements* to existing machines, substances, gadgets, or processes. A **design patent** protects the particular design of a product, including its shape, size, and appearance. Last, a **plant patent** protects the efforts involved with developing new plants, such as a hardier, drought-tolerant tomato plant.

If you invent something new and unique, it's important to know you must patent it to safeguard your intellectual property. But before you get started developing a new invention, you should make sure nobody else has come up with the same idea and patented it. Otherwise, you're going to waste a lot of time, money, and effort.

Finding Patents

Patents can be searched in several free, online databases, all of which can be found with a Google search:

- Patent Public Search (from the U.S. Patent and Trademark Office)
- PatentScope
- EspaceNet
- Lens Patent Search
- Google Patents

Some campus libraries provide additional databases for finding patents, such as:

- Derwent Innovations Index
- Bloomberg Law Patent Search
- WestLaw Edge

And, to find chemical and substance patents, your campus library may provide:

- Reaxys
- SciFinder

Consult the section in Chapter 20 on **content-specific databases** to learn where to find patent databases on your campus library's website or ask staff at your campus library for help locating these databases. If you need help searching, check to see if your library has a librarian with patent expertise. If your library doesn't have a patent expert, then search Google for: **USPTO patent search help**. This search will retrieve resources from the USPTO for getting started with patent searching.

An Overview of Standards

What's a standard? The **International Standards Organization (ISO)** provides an excellent definition: "A standard is a document that provides requirements, specifications, guidelines, or characteristics that can be used consistently to ensure that materials, products, processes and services are fit for their purpose." By the way, the ISO is an organization that coordinates the development of standards on an international level.

To understand how important standards are, imagine you're building a house and you've reached the point where you're putting in pipes for the plumbing. Suppose each pipe manufacturer make pipes in different sizes. That means you can't buy pipes from different manufacturers because they won't fit together. You could plumb your house with pipes from Manufacturer X, but they have a bit of a monopoly on the pipe market—and they know it—so they charge exorbitant prices for their pipes. But if you go with a smaller manufacturer, there's more risk they'll go out of business and then, in a few years' time, if you need to make repairs and replace a pipe, you'll have to replumb your entire house because replacement pipes won't be available from that manufacturer. Even if you decide to plumb your house with the overpriced pipes from Manufacturer X, there's a chance they could eventually go out of business too, so you're still faced with the prospect of having to replumb your entire house anytime you need to make plumbing repairs.

The existence of standards eliminates these issues. Standards dictate that certain products, like plumbing pipes, must all be of a certain quality, size, and shape no matter who manufactures them. Standards establish uniformity and interchangeability as well as ensuring that products (or methods) are high quality, reliable, and safe.

Standards are given a title that describes what the standard covers, and the title is usually preceded by an initialism identifying the society or agency that issued the standard, along with a number and date. Here are a few examples:

> ASTM C1225-08 (Reapproved 2016) Standard Specification for Fiber-Cement Roofing Shingles, Shakes, and Slates
>
> IEEE 1937.1 (2020) Interface Requirements and Performance Characteristics of Payload Devices in Drones
>
> ISO 21070 (2nd Edition, October 2017) Ships and marine technology—Marine environment protection—Management and handling of shipboard garbage
>
> ANSI/NISO Z39.19 (2005 Edition) Guidelines for the Construction, Format, and Management of Monolingual Controlled Vocabularies

Many engineers regularly refer to standards in their work. So, if you're planning on a career in engineering, knowing how to locate and read standards could be an essential skill for landing a job—and keeping that job.

Who Establishes and Publishes Standards?

There are a variety of different bodies that establish and publish standards. For example, certain societies and associations create and maintain standards, such as:

- ANSI—American National Standards Institute
- IEEE—Institute of Electrical and Electronics Engineers
- ASME—American Society of Mechanical Engineers
- ASCE—American Society of Civil Engineers
- NISO—National Information Standards Organization
- SAE—Society of Automotive Engineers

Standards are also developed by government agencies. In the United States such bodies include:

- OSHA—Occupational Safety and Health Administration
- ASSIST—U.S. Department of Defense Standards

There are also international organizations that establish standards, such as:

- ASTM-International—International Society for Testing and Materials
- ISO—International Organization for Standardization
- IEC—International Electrotechnical Commission

Finding Standards

If you need a specific standard or if you're browsing for relevant standards, then start your search on your campus library's website. Look for a search box on your library's homepage that invites you to search articles, books, media, and more, or has wording similar to that. This is called a **discovery tool** and you'll learn more about it in Chapter 20. You can enter a search for a specific standard, or you can enter keywords to look for standards that relate to a particular topic or issue.

If you're looking for a particular standard, you might want to search just the title but note you may need to add the initials of the issuing agency and/or the standard's number to zero in on the specific standard you need. Once you've run your search, look for filters (which are usually to the left of your results) to limit your results to standards only. (You saw how to do that for a book search in Figure 3.2.)

If your library doesn't provide a discovery tool search box on their homepage, or if the search you conducted didn't find what you needed, then you can ask staff at your library for help, or you can consult Chapter 20 for tips on navigating your library's website to find **content-specific databases** that specialize in providing standards, such as TechStreet or ASTM Compass. Some **publisher databases** also have lots of standards, such as IEEE Xplore or SAE Digital Library.

Other options for finding standards include searching Google for a particular standard to see if you can find and access it for free. Or go to the website for the society, organization, or agency that produced the standard and search for it there. You can also browse for standards using these two databases, both of which are easily found with a Google search:

- ANSI Webstore
- IHS Markit Standards Store

An Overview of Codes

A code is basically a standard or set of standards that are mandatory and enforceable by law. In the United States, national codes are documented in the **Code of Federal Regulations**. Figure 13.2 shows you the code for issuing special license plates for persons with disabilities taken from the "Highways" section of the Code of Federal Regulations.

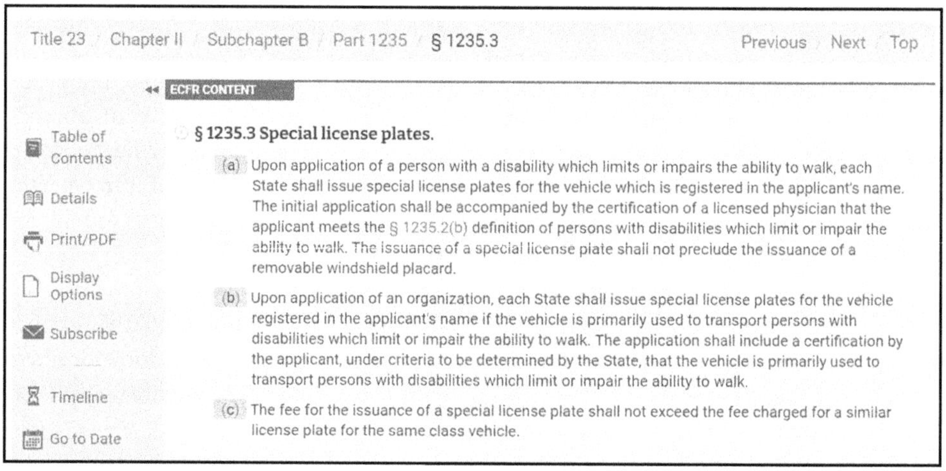

Title 23 / Chapter II / Subchapter B / Part 1235 / § 1235.3 Previous / Next / Top

ECFR CONTENT

Table of Contents
Details
Print/PDF
Display Options
Subscribe
Timeline
Go to Date

§ 1235.3 Special license plates.

(a) Upon application of a person with a disability which limits or impairs the ability to walk, each State shall issue special license plates for the vehicle which is registered in the applicant's name. The initial application shall be accompanied by the certification of a licensed physician that the applicant meets the § 1235.2(b) definition of persons with disabilities which limit or impair the ability to walk. The issuance of a special license plate shall not preclude the issuance of a removable windshield placard.

(b) Upon application of an organization, each State shall issue special license plates for the vehicle registered in the applicant's name if the vehicle is primarily used to transport persons with disabilities which limit or impair the ability to walk. The application shall include a certification by the applicant, under criteria to be determined by the State, that the vehicle is primarily used to transport persons with disabilities which limit or impair the ability to walk.

(c) The fee for the issuance of a special license plate shall not exceed the fee charged for a similar license plate for the same class vehicle.

Figure 13.2

Other important codes include building codes, fire codes, and electrical codes. In the United States, the **International Building Code** is used as the basis for most local city and county building codes, while the **National Fire Protection Association (NFPA)** establishes codes like these:

National Fire Protection Association (2019). NFPA—72 National Fire Alarm and Signaling.

National Fire Protection Association (2020). NFPA—70 National Electrical Code.

Finding Codes

If you need to find a specific code, then you can try searching for it on your library's website. Look for a search box on your library's homepage that invites you to search articles, books, media, and more—or has wording similar to that. This is called a **discovery tool** and you'll learn more about it in Chapter 20.

If your library doesn't provide a discovery tool search box on their homepage, or the search you conducted didn't find what you needed, then search Google to see if you can find a copy of the code for free. Or, go to the website for the society, organization, or agency that produced the code and search for it there—for example:

- The Code of Federal Regulations is searchable from the eCFR website ("eCFR" stands for "electronic Code of Federal Regulations").
- You can browse a list of National Fire Protection codes on the NFPA website.
- Go to the IAPMO website to look for codes developed by the International Association of Plumbing and Mechanical Officials.
- Go to the International Code Council (ICC) website to view their building safety I-Codes. (Or just do a Google search on **ICC I-Codes**.)

Some of the databases listed in the section above on "Finding Standards" will also have codes, so you can search those databases as well. If you need help finding or using any of those databases, then consult staff at your campus library.

To find local municipal building codes, try searching **municode.com**. You can browse to a particular city or town (a.k.a. municipality) by state, then look at particular codes by topic.

If you can't find the local codes you need from municode.com, then try a Google search on the city or town name along with the words "building codes" to see what you can find. (Obviously if you are not looking for building codes then use whatever search terms represent the type of code you want to find.) You can also go to a local government website for a city or town to see if the site offers a way to browse to the codes for that municipality.

Dealing with Access Problems for Patents, Standards, and Codes

If you can't find the patent, standard, or code you need, then your best option is to ask staff at your library for advice.

To sum up, what should science and engineering students know about patents, standards, and codes?

1. Patents protect your right to make money from your invention.
2. Patents can be searched in databases available on the internet and in certain library databases.
3. A standard is a publication that provides specifications for certain products, processes, or services.
4. There are a variety of ways to locate standards including searching library databases or using resources available on the internet.
5. A code is a standard or set of standards that are mandatory and enforceable by law.
6. You can search certain library databases for codes or use resources available on the internet.

What's Next?

You're finished with **Part II** of this book, which was all about different source types, their characteristics, and how to find them! Now we'll move on now to **Part III**, where you'll learn about key qualities sources have. We'll start by discussing the quality of being **scholarly**.

PART III
IMPORTANT QUALITIES OF SOURCES

CHAPTER 14
SCHOLARLY SOURCES

Scientists and engineers rely heavily on scholarly sources, so you'll need to know what these sources are and why they're important. In this chapter you'll learn about different types of scholarly sources and what makes them scholarly. This chapter will also explain the peer-review process that some scholarly sources undergo before being published. You'll also learn that scholarly, peer-reviewed sources are not perfect or infallible—nor are they the only place where authoritative information can be found.

The Terminology of Scholarly Sources

Before we dive into an explanation of scholarly sources, let's deal with some terminology issues. Recall from the chapter on journals (Chapter 6) the qualities "scholarly" and "academic" are two different ways of saying the same thing. In this book we're using "scholarly" as the preferred term and in Figure 14.1 you see how "peer-reviewed sources" are a type—or subset—of scholarly sources.

Thus, "scholarly" is an umbrella term that includes sources that are peer-reviewed, and we can just ditch the term "academic" because it means the same thing as "scholarly." But be aware that your instructors may have different ideas about how these terms relate to each other *and* how these terms should be defined. If your instructors mention source qualities like scholarly, academic, peer-reviewed, etc., in a lecture or in an assignment, you should clarify with them how they define these terms. For more advice on why and how to do that, see Chapter 18. With that settled, let's talk about scholarly sources in detail.

An Overview of Scholarly Sources

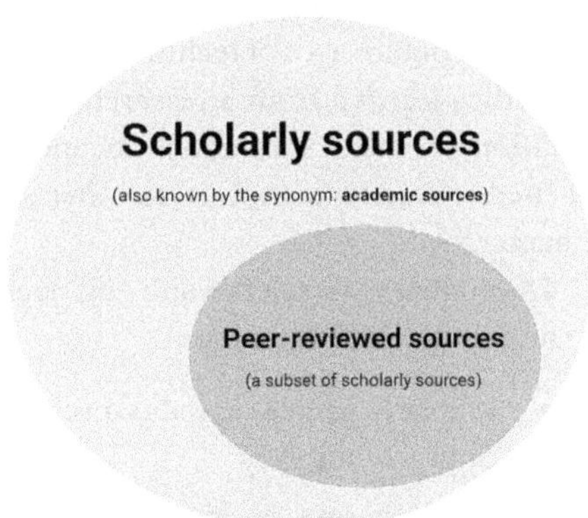

Figure 14.1

Scholarly sources present the results of scholarship that was conducted by scholars. That's a pretty circular definition, so let's break it down into smaller chunks. **Scholars** are experts with extensive knowledge of a particular subject area. Scholars usually have advanced educational degrees and/or other professional training, as well as experience conducting research on their specialty topic. So, scholarly sources are written by people who know what they are talking about. Really, you can lump almost any professional researcher into the category of "scholar" as long as they have relevant credentials or expertise, as well as research experience.

Scholarship is the act of **conducting research** on a topic—and not just any topic, but rather a scholarly topic. A **scholarly topic** is any topic relating to disciplines like biology, medicine, sociology, or engineering—in fact, all the disciplines that are represented by academic departments on a college campus. For example, bioengineering is a scholarly discipline, and an example of scholarship in that discipline would be conducting research on the optimal design of prosthetic limbs. This research—like *all* scholarly research in science and engineering disciplines—would lead to new discoveries, facts, and technological applications.

Meanwhile, countless other topics are not considered scholarly. For these topics, there's no reason for experts to study them intently, nor would research on these topics result in groundbreaking new facts and applications. Examples of non-scholarly topics include:

- traveling Europe cheaply
- starting a knitting club
- doing yoga at home
- making lures for trout fishing

The reason scholars publish the results of their scholarship is to share those results with other scholars so they can all learn from each other and build upon each other's research. So, scholarly sources are written *by* scholars *for* other scholars. Scholarly sources are also used by college and graduate students who want to gain a better understanding of a particular area of research. Plus, some types of policy-makers and managers consult scholarly sources when they need to make well-informed decisions. For example, a government environmental manager would likely consult scholarly sources when deciding on the best way to sustainably manage a forest or river.

The writing style for scholarly sources is usually formal, neutral, and unemotional, and may contain a lot of technical terminology. Furthermore, when scholars write about what they learned from a research project, they typically **cite** other sources to provide additional context, substantiation, and evidence. So scholarly sources often contain a lot of cited references compared to other types of sources. You'll learn more about citing in Chapter 29.

In summary, for science and engineering disciplines, these are the qualities we expect to find in scholarly sources:

- scholarly sources are about scholarly topics (like geology, electrical engineering, or psychology)
- scholarly sources describe research
- scholarly sources are written by scholars

- scholarly sources are characterized by a serious, unemotional, and technical writing style that's obviously directed at other scholars, not the general public
- scholarly sources usually contain citations to other sources (also known as "references")
- scholarly sources undergo peer review and/or editorial review before being published (more on this below)

The Whole Point of the "Scholarly" Label

In previous chapters you learned about different source types and how they vary in content and purpose to serve different information needs. This knowledge will go a long way toward helping you zero in on the best type of source to resolve a particular information need. But sometimes it's not quite enough. Often you need information that has a certain set of *qualities*. In your career as a scientist or engineer, there will probably be many occasions when you need to consult research information that's been written by a knowledgeable expert/scholar and reviewed for accuracy prior to publication. The label "scholarly" solves that problem by leading you to sources with those particular qualities.

Types of Sources That Are Considered Scholarly

Most scholarship is published in the following source types:

- **research articles** and **review articles** found in **journals**
- **conference papers** published in **conference proceedings**
- **books** (but if you recall from Chapter 3, there are different types of books and only some qualify as scholarly)

Do you recall from Chapter 6 that many journals contain content besides research articles and review articles? Some of that other content may also be considered scholarly, such as perspective articles, case records, research analyses, technical notes, etc. For example, below you see a **perspective article** from the journal *Nature* that would be considered scholarly:

> Joseph, D., et al. (2022). Transitioning organizations to post-quantum cryptography. *Nature*, *605*(7909): 237–243.

In this article, the authors present their perspectives and recommendations on how to move organizations from pre-quantum cryptography to post-quantum cryptography (PQC), including "transition timelines, leading strategies to protect systems against quantum attacks, and approaches for combining pre-quantum cryptography with PQC to minimize transition risks." The authors base their recommendations on a review of existing published research, and they cite that research to further support their claims. So, this article qualifies as a scholarly source.

Many journals also contain editorials, letters, news, and the like. Although this sort of content *supports* scholarship, whether it should be labeled "scholarly" is debatable. Some people would say "yes," whereas others would say "no"—it really depends on how strictly you define the word "scholarly."

Determining if a Source is Authored by a Scholar

One of the characteristics of scholarly sources is that they're written by scholars. But when you look at a source, how do you know if the authors are qualified scholars, researchers, or other experts? Check the source to see if the authors' educational degrees are provided after their names. Sometimes they are and sometimes they aren't—but if they are, look for graduate-level degrees like PhD, MD, MS, MA, etc., because those degrees indicate advanced training.

You can also look for clues about where the authors are employed. Most scholars work for employers who support and promote research efforts. Such employers include colleges, universities, research institutes, research organizations, as well as certain government agencies and some companies. If the source indicates that the authors hold jobs with one of these types of employers, then you can assume they are scholars or experts, because those types of employers hire experts. Some sources (like journal articles and conference papers) explicitly state the author's place of employment. Take a look at the journal article shown in Figure 7.1. Just under the title ("Predicted takeover responses to silent automated vehicle failures") you see the names of the authors, and below their names are the institutions where they work—showing they are all employed by universities.

If the source doesn't provide much information about the author, then you can conduct a Google search on their name to see what you can learn about their educational degrees, employer, or other credentials.

The Review Process for Scholarly Sources

Before being published, scholarly sources undergo a review process. The review process begins when an author sends their **manuscript** off to a publisher in the hope it'll be accepted for publication. What happens next varies, but usually an **editor** who works for the publisher looks over the manuscript to determine if it's good enough to review in more detail, or if it should be rejected. Manuscripts get rejected for lots of reasons, but often it's because the manuscript is poorly written or because the topic of the manuscript falls outside the scope of what that publisher typically publishes. If the editor decides the manuscript is

worth examining more closely, then it will undergo **editorial review** or **peer review**—and some publications undergo *both* editorial review and peer review.

Editorial review is usually performed by an editor employed by the publisher. The editor will work with the author to improve the readability of the manuscript to ensure it flows, is coherent, succinct, and understandable. The editor may also look at the content itself to assess whether the research appears to be rigorous and sound. Or, the editor may bring in outside scholars or experts who have expertise on the manuscript's subject matter, so those experts can weigh in on the quality and rigor of the manuscript's content. These outside experts are called **peer reviewers** and you'll learn more about them in a moment.

After the editorial and/or peer-review processes are complete, the manuscript will undergo **copy-editing**, which is performed by a **copyeditor** who works for the publisher. The copyeditor corrects any grammatical errors in the manuscript. Once the copyeditor is finished, the manuscript is ready for publication, and—as you know from previous chapters—it might get published in print, online, or both.

The Peer-Review Process in More Detail

Many scholarly sources undergo **peer review** before being published. The peer-review process is organized by an editor who works for the publisher. The editor will identify a few scholars (often three) who specialize in the same discipline as the author of the manuscript. These scholars are likely to be faculty at a college or university, or they might be employed by some other agency or organization that conducts research. The point is they are **peers** of the author because they're all experts working in the same discipline. The editor will send these peers the manuscript and ask them to review it. If the peers are not too busy and agree to do it, they become **peer reviewers**.

Peer reviewers will examine the manuscript for quality and rigor and try to catch any mistakes. They'll focus on whether the study design and research methods are robust, and whether there is appropriate evidence to support the author's conclusions. They might also critique the importance and utility of the research, the organization and readability of the author's writing, and they may even correct grammatical mistakes they notice. If the peer reviewers are unsatisfied with the manuscript, they'll let the editor know they think the manuscript should be rejected and not published. Alternatively, they may find the manuscript mostly satisfactory but in need of revisions. They'll write up their recommended revisions, which will be passed along to the author through the editor. If the author decides to resubmit the manuscript, they must address the critiques made by the peer reviewers. Then, the editor will work with the peer reviewers to evaluate the revised manuscript to see if all the problems have been fixed. If not, the manuscript may go back to the author to make more changes. In fact, there might be several rounds of revisions before the editor and peer reviewers are satisfied the manuscript is ready for publication. This process may take a few weeks but usually it takes months. To see how peer review fits in the research cycle, take a look at the diagram in Figure 14.2.

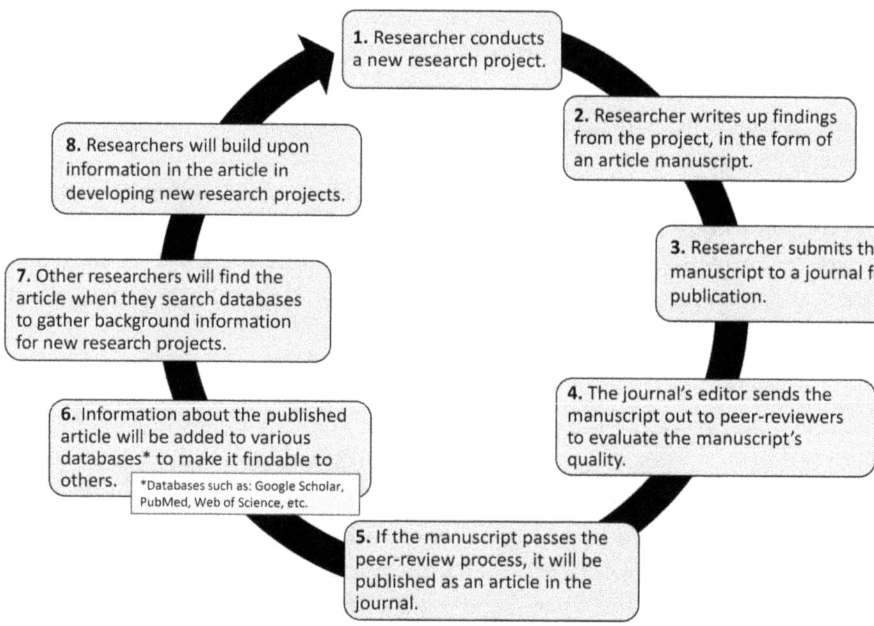

Figure 14.2

Just so you know, sometimes you hear peer reviewers called **referees**. In sports, referees make sure the rules of the game are followed and they settle disputes, but within scholarship, referees evaluate manuscripts and decide whether the manuscript is good enough to publish. So, a referee is the exact same thing as a peer reviewer. Likewise, you may hear people refer to peer-reviewed sources as **refereed sources**. Both mean the same thing.

During the peer-review process, the author usually isn't told who their peer reviewers are. This is called **anonymous peer review** or **blind peer review**. The purpose of anonymity is to allow the peer reviewers to say what they think about the manuscript without potentially upsetting the author. Because both the author and the peer reviewers are experts in the same discipline, they may know each other and be colleagues or friends. Thus, anonymity allows peer reviewers to freely give comments without the risk of offending the author. Of course, not all scholars working in the same field are friendly with each other, so some publishers employ **double-anonymous peer review** (also called **double-blind peer review** or **masked peer review**) where the author doesn't know who the peer reviewers are, nor do the peer reviewers know who the author is. This helps prevent peer reviewers from being unreasonably harsh in their evaluation of the manuscript just because they happen to dislike the author.

On the other hand, some publishers have opted for a peer-review process that's completely transparent (called **open peer review**), where the author and the peer reviewers know who each other are.

The Value of Peer Review

You may recall from Chapter 1 that new discoveries and inventions don't just emerge from nothing—they come about when researchers build upon information originating from earlier discoveries. To illustrate that point, we noted that the journal article below builds upon research that already established that nitrogen is a damaging pollutant.

> Craig, L. S., et al. (2008). Stream restoration strategies for reducing river nitrogen loads. *Frontiers in Ecology and the Environment*, *6*(10): 529–538.

Suppose you conducted research on how to reduce the amount of nitrogen in streams and other aquatic ecosystems, then later learned that all the earlier research showing that nitrogen is a pollutant was wrong; rather, it causes no damage to aquatic ecosystems. What a waste of your time and salary as well as the time and salaries of other researchers who were working on how to *remove* nitrogen from streams!

You can see from that example the stakes can be high to get information *right*—especially when others will be referring to that information and using it as a foundation for their research projects. So, the goal of the peer-review process is to improve the quality, accuracy, and reliability of research that gets published, because other researchers will be building upon that published research.

Problems with the Peer-Review Process

There are a variety of problems with peer review, as well as controversies about how effective it is. In this section we'll look at a few problems, but if you want to learn more, just conduct the following Google search:

> **peer review (problems OR flaws OR limitations OR disadvantages OR drawbacks OR criticisms)**

One problem is that peer-review might not be very effective. Several studies of the peer-review process suggest it only catches 25–30 percent of errors in a manuscript.[1] This may be because peer reviewers don't always have access to the raw data associated with the research project, nor do they have all the details about how those data were collected and analyzed. However, even if they *did* have this information, they'd need a lot of time to review it and most peer reviewers simply don't have time. Thus, they are constrained by how thoroughly they can assess a manuscript. That means the process of peer review may be giving us a false sense of security that most published research is quite rigorous when it's not.

Another problem is that unresolved philosophical disputes among authors and reviewers can hijack the peer-review process. For example, researchers in taxonomy tend to be divided into two camps: splitters and lumpers. Splitters are interested in differences among organisms and are quick to classify a group of organisms as its own species. Lumpers do just the opposite—they are more inclined to dismiss differences and recognize fewer species. Now imagine a combination of lumpers and splitters peer-reviewing a manuscript written by a splitter that's about splitting up a currently recognized species into three new species. The peer reviewers who are lumpers will want to reject the manuscript because they hold a different view on this issue. The peer reviewers who are splitters might think the manuscript is great.

A third problem is that bias still creeps into the peer-review process. Remember earlier in this chapter you read about how peer reviewers might be biased in some way toward the author of the manuscript they're reviewing? Although publishers attempt to reduce this bias through procedures such as double-blind peer review, it's impossible to entirely eliminate it. Often reviewers and authors can guess who each other are and this

---

[1]See for example:

1) Baxt, W. G., et al. (1998). Who reviews the reviewers? Feasibility of using a fictitious manuscript to evaluate peer reviewer performance. *Annals of Emergency Medicine*, *32*(3): 310–317
2) Schroter, S., et al. (2008). What errors do peer reviewers detect, and does training improve their ability to detect them?. *Journal of the Royal Society of Medicine*, *101*(10): 507–514.
3) Godlee, F., et al. (1998). Effect on the quality of peer review of blinding reviewers and asking them to sign their reports: a randomized controlled trial. *JAMA*, *280*(3): 237–240.

can lead to a reviewer being unduly harsh or too lenient on a manuscript—depending on their relationship with the manuscript's author. When a peer reviewer is not reviewing a manuscript objectively, the quality of the manuscript in its final, published form could be negatively affected.

Which Scholarly Sources Are Peer-Reviewed and How Can You Tell?

In general, journal articles—including both research articles and review articles—undergo a peer-review process before being published. If you want to make absolutely sure that a particular journal article is peer-reviewed, you'll need to investigate whether the journal it was published in utilizes a peer-review process. To do that, search the journal's title in Google to find its website. Once there, you might find information about whether it's a peer-reviewed journal right on the homepage. Or you may have to poke around a bit and look for links that lead to instructions on how to submit manuscripts to that journal (see Figure 14.3). Within those instructions, there's usually information about the peer-review process that submitted manuscripts undergo, thus confirming the journal has a peer-review process in place.

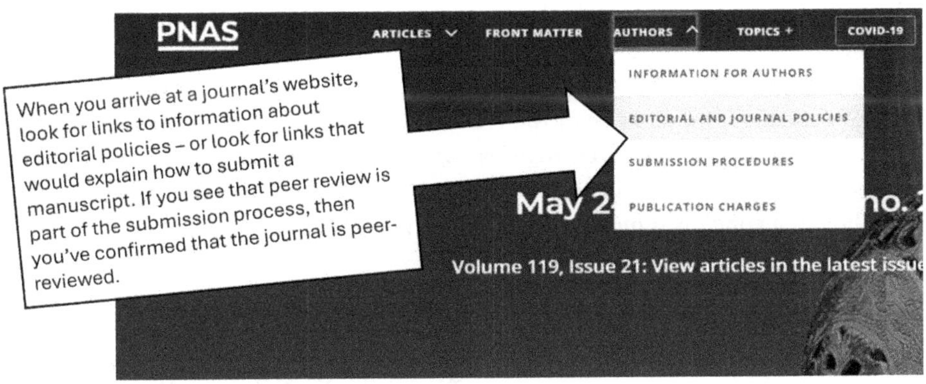

Figure 14.3

Sometimes it's even quicker to run a Google search on the title of the journal along with the words "peer review." You may immediately see evidence of peer review within your results, as you do in the first two Google results shown in Figure 14.4.

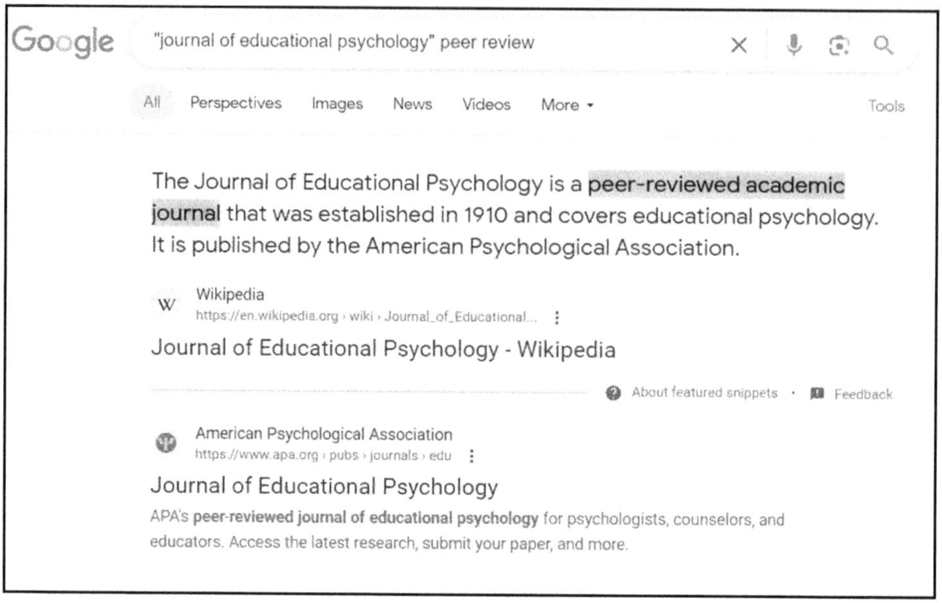

Figure 14.4

Research articles and review articles are always peer-reviewed, but there may be other content within a journal that's *not* peer-reviewed. For example, if a journal offers content such as editorials, correspondence, opinions, etc., then that content is not peer-reviewed because it's simply offering a point of view. In Figure 14.5 you see a typical table of contents from an issue of the journal *Nature Ecology and Evolution*. Only the "Review Articles" and "Research Articles" will undergo a peer-review process prior to publication. However, we still refer to the entire journal as a "peer-reviewed journal," even though it contains some content that's not peer-reviewed.

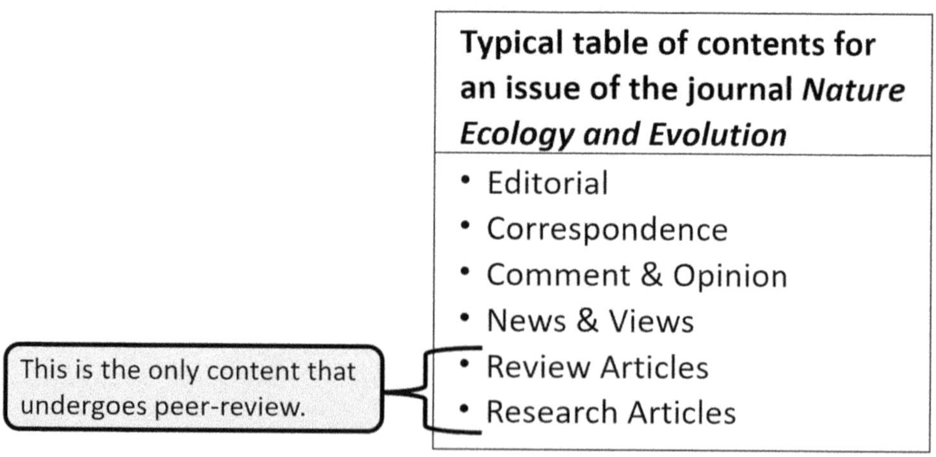

Figure 14.5

If you run into difficulties determining whether a journal is peer-reviewed—or whether a specific article or other content within a journal is peer-reviewed—then ask library staff for assistance.

Now let's move on to conference papers. Typically, conference papers are either peer-reviewed or editorially reviewed (and sometimes both), but often it's not clear what type of review process they've undergone or how rigorous it was. Even if a conference paper *is* peer-reviewed, the process may be more relaxed than for a journal article—for example, there may be only *one* peer reviewer rather than three.

You can look for information about the peer-review process on the conference publisher's website, but you may not find anything. If you don't, that doesn't mean the conference paper wasn't peer-reviewed. It may just be that the publisher hasn't bothered to indicate one way or another. So, you may not be able to confirm whether a specific conference paper was peer-reviewed. Of course, you could try contacting either the publisher and/or the author of the paper to inquire, if you really need to know.

Scholarly books are similar—they might be peer-reviewed or editorially reviewed (or both). Like conference papers, you can look for peer-review information on the book publisher's website, but you may not find anything. Again, that doesn't mean the book was not peer-reviewed. If you really need to know for sure, then you may need to contact the publisher and/or author to confirm.

In fact, journal articles are the only source for which peer review is both a standard procedure as well as fairly easy to verify—and this is why you might encounter people who consider journal articles to be the *only* type of source that's peer-reviewed. They're not—they're just the only type of source that consistently uses peer review as an editorial process and for which information about the peer-review process is transparent and easy to confirm.

You should now have a better understanding of the peer-review process, including how it works, why it's important, which sources undergo peer review, and how to determine if a particular source is peer-reviewed. So, let's get back to talking about scholarly sources in general—which, as you now know—include peer-reviewed sources.

Should You *Only* Use Scholarly Sources?

There's a common misconception among students that scholarly sources are the *only* sources you should use because they're written by scholars and carefully reviewed—thus they're the most authoritative source. But whether a scholarly source is the best source to consult really depends on your particular information need. For example, if you want to know what the weather will be tomorrow, a scholarly source is a terrible place to look for that information. So, scholarly sources are not always the best or most appropriate source to consult.

Furthermore, in certain disciplines like ecology, traditional ecological knowledge can be an extremely important source of information, yet it's not considered scholarly.

Traditional ecological knowledge (TEK) refers to the knowledge and skills gained from hundreds or thousands of years of direct experience with the local environment passed down within indigenous communities from generation to generation.

Let's look an example of the value of TEK. At one time the conventional wisdom among professional foresters was that forest fires should be prevented. So, research efforts were directed at how best to do that, and policies were put into place that successfully reduced the number of forest fires. However, forest fires occur naturally, so preventing them results in disruptions to natural ecological processes—with one result being denser forests. When denser forests catch fire, they are harder to put out and they burn hotter and more severely, causing damage to the ecosystem.

It didn't take long for forestry scholars to realize that preventing fires causes its own set of problems and to notice that traditional ecological knowledge offers a better model of forest fire management. Indigenous communities managed fire using practices other than widespread fire suppression and had better ecological outcomes. In fact, below you see some examples of scholarly sources that discuss the merits of *non-scholarly knowledge* in forest fire management!

Lake, F. K., et al. (2017). Returning fire to the land: celebrating traditional knowledge and fire. *Journal of Forestry*, *115*(5): 343–353.

Mason, L., et al. (2012). Listening and learning from traditional knowledge and Western science: a dialogue on contemporary challenges of forest health and wildfire. *Journal of Forestry*, *110*(4): 187–193.

Kimmerer, R. W. & Lake, F. K. (2001). The role of indigenous burning in land management. *Journal of Forestry*, *99*(11): 36–41.

So scholarly publications are one source of knowledge and authority, but to be truly information literate, one must "view authority with an attitude of informed skepticism and an openness to new perspectives, additional voices, and changes in schools of thought."[2]

However, it's also true that in college and graduate school the use of scholarly sources is heavily emphasized—in fact, you may hear your instructors tell you to *only* use scholarly (or peer-reviewed) sources. Often that's because instructors aren't confident you've gained the information literacy skills to judge the credibility of other types of sources. And instructors are right, most students are still developing their ability to distinguish credible sources from those that are not. Don't worry, in Chapter 17 you'll learn how to evaluate whether a source is credible. But for now, just know that *all* scholarly sources are considered to be credible sources and that's usually why instructors insist you use them.

---

[2]Framework for Information Literacy for Higher Education (2015). American Library Association. http://www.ala.org/acrl/standards/ilframework (Accessed March 13, 2023).

The problem is, when students hear their instructors urge them to use scholarly sources, many start to believe they should *only ever* use scholarly sources. That's not the case. Instead, you should always strive to use the most appropriate source for your particular information need.

Scholarly Sources Are Imperfect

We have high standards for scholarly sources insofar as we expect them to be written by qualified experts who know what they are talking about. Plus, we expect the content in these sources to be edited and reviewed by other experts to ensure it's rigorous before the source gets published. Because of these standards, scholarly sources represent *our current best effort* to share reliable and accurate information. But in the end, it's just an *effort*, it's not a guarantee, which means that information in scholarly sources can be flawed and unreliable.

There are many factors that work against our best efforts to make scholarly sources as high-quality and reliable as possible, so let's review some of them. For one thing, the "publish or perish" requirement interferes with the quality of scholarly sources in a variety of ways. For example, this requirement means that researchers are incentivized to publish as many scholarly sources as possible. As you can imagine, that's not conducive to careful, deliberate research, plus it may cause researchers to avoid certain research projects altogether, if they think those projects won't lead to lots of publications. Also, the publish or perish requirement, combined with the fact that publishers don't like publishing negative results, means there's an incentive for researchers to exaggerate how positive their results are. (For more on this issue, see Chapter 7 and Appendix B.)

Another major problem is that researchers—like all human beings—get caught up in politics, ideologies, and tribalism, and this can negatively affect their ability to be impartial. Naturally this can impact the quality and integrity of the research they publish in scholarly sources. Also, researchers simply make mistakes sometimes. There's lots of room for error in the design of research projects *and* when interpreting results. Nor do peer reviewers always catch these mistakes. Take, for example, this article:

> AlShebli, B., et al. (2020). The association between early career informal mentorship in academic collaborations and junior author performance. *Nature Communications* 11(1): 5855.

The researchers who authored this study conducted a statistical analysis that indicated junior women scientists were more successful when their mentors were male rather than female. These researchers concluded that whereas "current diversity policies encourage same-gender mentorships to retain women in academia, our findings raise the possibility that opposite-gender mentorship may actually increase the impact of women who pursue a scientific career."

Shortly after this article was published, many scientists contacted the publisher to say they disagreed with the methods these authors used and the assumptions the authors made about the sorts of relationships that constitute "mentorship." In other words, the validity of the study, including how it was conducted and how the results were interpreted, was called into question. After reviewing the study, both the publisher and authors agreed that the study was flawed.

Another reason scholarly sources contain imperfect information is because research in science and engineering is never really done—rather, it's always in a state of progress. While each discipline has its established facts and theories, there are also many uncertainties and unknowns—especially at the cutting edge of research. In fact, it's not uncommon for researchers to conduct studies that result in conflicting results and conclusions, and it's only over time that enough evidence emerges, from multiple published research studies, to support one conclusion over another. The more evidence we have that supports a particular research finding, and the longer that evidence remains unrefuted, the more confident we can be that the research finding is an actual fact and no longer just a tentative conclusion.

However, sometimes even well-established facts get revised, updated, or even discarded, as new evidence emerges. So, when you consult a scholarly source, you are seeing a snapshot of what we know (or think we know) at a particular time. However, that knowledge may get revised and updated in the future, as we collect more evidence.

The emerging field of **metascience** looks at how the research process works and how it could be improved to be more efficient, as well as more accurate and reliable. As progress is made in this field, we may see changes in the way research is conducted, funded, vetted, and incentivized that will improve the overall reliability and accuracy of scholarly sources.

Formal Retractions of Flawed Information

When scientists and engineers notice flaws in published scholarly information, they can report those flaws to the publisher. Usually, the publisher will investigate the issue and if it's determined the flaws are serious, the publisher will retract the publication. To **retract** a publication means to remove the article or paper from the publisher's website, or leave it up but with a notification to the reader that the publication has been retracted. The publisher may also issue a **retraction notice** that provides reasons for the retraction. Figure 14.6 shows you the article discussed in the previous section. This article was retracted but left up on the publisher's website with a banner showing it's been retracted. For up-to-date news on scholarly content that's been retracted, consult **retractionwatch.com**.

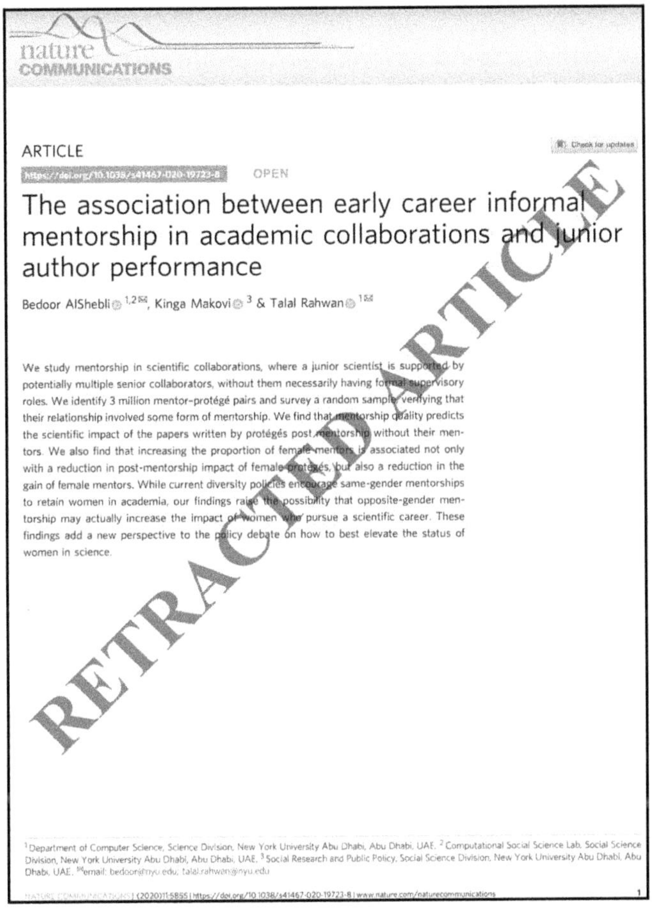

nature COMMUNICATIONS

ARTICLE

Check for updates

https://doi.org/10.1038/s41467-020-19723-8 OPEN

The association between early career informal mentorship in academic collaborations and junior author performance

Bedoor AlShebli[1,2], Kinga Makovi[3] & Talal Rahwan[1]

We study mentorship in scientific collaborations, where a junior scientist is supported by potentially multiple senior collaborators, without them necessarily having formal supervisory roles. We identify 3 million mentor-protégé pairs and survey a random sample, verifying that their relationship involved some form of mentorship. We find that mentorship quality predicts the scientific impact of the papers written by protégés post mentorship without their mentors. We also find that increasing the proportion of female mentors is associated not only with a reduction in post-mentorship impact of female protégés, but also a reduction in the gain of female mentors. While current diversity policies encourage same-gender mentorships to retain women in academia, our findings raise the possibility that opposite-gender mentorship may actually increase the impact of women who pursue a scientific career. These findings add a new perspective to the policy debate on how to best elevate the status of women in science.

RETRACTED ARTICLE

[1] Department of Computer Science, Science Division, New York University Abu Dhabi, Abu Dhabi, UAE. [2] Computational Social Science Lab, Social Science Division, New York University Abu Dhabi, Abu Dhabi, UAE. [3] Social Research and Public Policy, Social Science Division, New York University Abu Dhabi, Abu Dhabi, UAE. email: bedoor@nyu.edu; talal.rahwan@nyu.edu

NATURE COMMUNICATIONS | (2020)11:5855 | https://doi.org/10.1038/s41467-020-19723-8 | www.nature.com/naturecommunications 1

Figure 14.6

To sum up, what should science and engineering students know about scholarly sources?

1. Your instructors may use different definitions for "scholarly," "academic," and "peer-reviewed" than how those words are defined in this book.
2. Scholarly sources describe research written by expert scholars, on scholarly topics.
3. Example of scholarly sources include journal articles, conference papers, and some books.
4. Scholarly sources undergo editorial review, peer review, or both. The review process helps ensure the source contains accurate and reliable information.
5. The peer-review process reduces the amount of flawed information that gets published but doesn't eliminate it. If a source gets published but then is found to contain flaws, it may get retracted.
6. Facts about the world derived from scholarly sources may get updated or revised over time as new evidence emerges.

What's Next?

This was a long chapter, but you learned a lot about scholarly sources and peer review. The next chapter will be much shorter and will focus on sources that are not scholarly, which we commonly refer to as **popular** sources.

CHAPTER 15
POPULAR SOURCES

You might encounter assignments that require you to find and use popular sources—or you might be told to *avoid* using popular sources. Either way, it'll be hard to follow those assignment instructions if you don't know what popular sources are. This chapter will explain what is commonly meant by "popular sources," as well as their good and bad qualities, and how these sources are used.

An Overview of Popular Sources

Popular sources fulfill countless information needs that are not scholarly in nature, such as providing news, supplying quick facts and information, supporting hobbies, showing how to do things, providing entertainment, etc. For example, suppose you want to find out the results of a recent election, or you want to find out what a hellbender is, or you need to learn how to repair a light fixture. Respectively, you might consult a newspaper article, a website like Wikipedia, or a YouTube video, all of which are considered popular sources. Below is a list of sources that are typically considered to be popular:

- Newspapers
- Magazines
- Websites (meaning *website-sources*)
- Non-scholarly books
- Media such as videos, podcasts, TV shows, etc.

Unlike scholarly sources, the audience for popular sources is the general public, thus these sources are much easier to read and understand.

Editorial Review of Popular Sources

You learned about the review process for scholarly sources in the last chapter, but what sort of review do popular sources undergo? It varies. There may be some editorial reviewing for readability, clarity, and grammar—and some popular sources are fact-checked. But usually there's no information available about the review process or how much care was taken to ensure the content is accurate and reliable. And for many popular sources, there's no review process whatsoever. A good example would be a personal blog where the author is writing directly to their reader as well as doing all their own editing—nobody else is involved in the process.

Not All Sources Fall Neatly into a Scholarly or Popular Category

Sometimes you hear people talk about scholarly and popular sources as if all types of sources fall into one of these two categories. But that's not the case. For example, dissertations and theses fit most of the criteria for being scholarly, except that the authors are students and so aren't considered expert scholars yet. But if you disqualify theses and dissertations from being scholarly, that certainly does not make them popular sources.

Next, consider textbooks. Many textbooks are written at a level that could be understood by a general audience (which is a characteristic of popular sources) but their purpose is to teach a scholarly topic in depth, which is not typical of popular sources. Also, trade magazines aren't scholarly sources, but because they're directed at a very specific, expert audience, they aren't popular sources either. Then you have sources like data, patents, standards, and codes for which we just don't tend to apply labels like "scholarly" or "popular." Table 15.1 shows that lots of sources don't fit into the scholarly or popular dichotomy.

Table 15.1

Typical examples of scholarly sources	*Typical examples of popular sources*	*Sources that don't cleanly fit as either popular or scholarly*
Journal articles Conference papers Scholarly books	Newspaper article Magazine article Websites Popular books Media (videos, podcasts, TV shows, etc.)	Dissertations Theses Textbooks Reference books Trade magazines Data Patents Standards Codes

Also, some sources have a mix of scholarly and popular qualities. Consider, for example, the website for the IUCN Red List of Threatened Species, which provides details on threatened species such as the bush dog. Here's the page discussing bush dogs (*Speothos venaticus*):

> DeMatteo, K., et al. (2011). *Speothos venaticus*. The IUCN Red List of Threatened Species 2011: e.T20468A9203243. https://dx.doi.org/10.2305/IUCN.UK.2011-2.RLTS.T20468A9203243.en.

If you take a look at that page, you'll see many different categories of information you can explore, such as Taxonomy, Geographic Range, Habitat and Ecology, Threats, Use and Trade, etc. Notice that a Bibliography is among those categories, and it provides all the references cited throughout this entry for the threatened bush dog. The presence of a bibliography (also called **cited references**) is a quality that's typical of scholarly sources. But this website has even more qualities that are typical of scholarly sources. For example, for each species account on this website the authors are listed and if you Google those authors to learn more about them, you'll find they are expert scholars working at universities or other types of research institutions. Plus, each species account is peer-reviewed, and the peer reviewers are listed.

So basically, the IUCN Red List of Threatened Species is a scholarly, peer-reviewed website that functions like an online reference book. Here you have a source that doesn't fall neatly into *any* obvious category. Don't be surprised if you come across other enigmatic (but extremely useful) sources like this one. And don't shy away from using a source just because it's hard to categorize.

In fact, this is as good a time as any to share with you that novice researchers (such as students) tend to be more comfortable thinking of sources as separated into distinct categories with thick lines between those categories. Books are books, articles are articles, scholarly sources are never popular, popular sources are never scholarly, etc. When you're a novice, these categories help you grasp the array of sources out there, and how they're designed for different purposes. But over time you'll find that lots of sources don't fit neatly into these categories. That's ok, though. Once you have the basics down, you'll be more at ease when you encounter a source that doesn't fit into the normal categories you've grown familiar with. You'll learn to examine that source and evaluate its content, purpose, and credibility on its own merits, and decide whether it's suitable for your particular information need.

Scholarly and Popular Sources Are Not on Opposite Ends of a Good–Bad Spectrum

You may recall from the last chapter that some students develop a misconception that they should *only* use scholarly sources, because those sources are written by experts and carefully reviewed. The obvious corollary to this is that popular sources are bad and shouldn't be used. This idea gets reinforced when students hear their instructors tell them *not* to consult and use popular sources for an assignment. But the reason instructors prefer that you use scholarly sources is because those sources are known to be credible, not because all popular sources are bad. In fact, popular sources are sometimes used for scholarly purposes! As an example, take a look at the journal article shown below. In this article, scholars gathered and analyzed information from a popular source (newspapers) to examine how climate change is portrayed in the media. Then, they described their findings in a journal article. So, they transformed information found in popular sources into information that is scholarly.

> Ford, J. D. & King, D. (2015). Coverage and framing of climate change adaptation in the media: a review of influential North American newspapers during 1993–2013. *Environmental Science and Policy*, *48*: 137–146.

It's true that some popular sources contain misinformation (and you'll learn more about this in the next section), but many contain excellent information. So, a better way to think about the categories "scholarly" and "popular" is that they serve different information needs—not that one is good and the other is bad.

Using Popular Sources Sensibly

You just learned that scholarly sources are not competing with popular sources to be superior; rather, these two types of sources are simply intended for different purposes. While this is true, you should still exercise caution using popular sources, as the quality of information found in these sources can be highly variable and range from excellent to terrible.

Suppose, for example, you're writing a research paper about the increasing public demand for companies to use more environmentally sustainable practices. In the course of your research, you come across a popular source—the website of an oil company. The website has lots of information about what the company is doing to be more environmentally sustainable and mitigate their impact on the environment.

Then suppose you found a journal article written by a team of ecology and environmental engineering faculty at a university. The journal article describes how these researchers investigated claims made by oil companies about their environmental sustainability practices, measured against the *actual* practices the companies engaged in. They found that oil companies (including the same oil company whose website you just looked at) greatly exaggerate their claims to be sustainable and deliberately avoid opportunities to mitigate their effects on the environment.

Which source would you believe? The **claims** made on the company's website, or the **evidence** provided by the study the scholars did? Obviously, you should be wary of the claims made on the company's website. It's in the company's interest to boost their image by stretching the truth to the extent they think they can get away with it. So, here's a case where you need to be very careful about information provided by a popular source.

But now let's consider another popular source, the Internet Movie Database (IMDb.com). This website provides information about who starred in various movies and TV shows, along with other bits of related trivia. It's a heavily used popular source, and people seem to instinctively trust it, but why is that? For one thing there's simply no reason to present the content in a way that's biased, exaggerated, or false. Also, if it was full of errors, then over time people would get fed up with it and stop using it, and it would probably get taken down. So, the type of content it provides, combined with its longevity, speaks to its value, utility, and credibility.

People may not be consciously processing reasons why they instinctively trust a popular source like IMDb and instinctively distrust claims made on a company's website about how exemplary that company is. But, when you think it through, there are good reasons for trusting one and not the other. You'll learn more skills to help you determine what—or what not—to trust in Chapter 17.

To sum up, what should science and engineering students know about popular sources?

1. Popular sources are characterized by being easy to understand and of interest to lots of people.
2. Not all types of sources can be categorized as either scholarly or popular.
3. It's not accurate to say that scholarly sources are good, while popular sources are bad; rather, these sources serve different purposes.
4. Exercise caution when using popular sources because the information they provide ranges in quality from excellent to terrible.

What's Next?

In the next chapter, you'll learn about the qualities **primary** and **secondary** and why they're important to scientists and engineers.

CHAPTER 16
PRIMARY AND SECONDARY SOURCES

It's likely you'll encounter assignments that ask you to find and use primary sources, secondary sources, or a combination of both. That'll be tricky, though, if you don't understand what "primary" and "secondary" mean or the qualities they signify, which this chapter will explain. You'll be even more confused if you don't realize that what counts as a primary or secondary source depends on the discipline. The types of sources considered primary or secondary in science and engineering *are not the same* as those that the humanities (such as art and literature) or some social sciences (such as history and law) would consider primary or secondary. This is important because your first encounter with these terms in the classroom may be from an instructor who defines "primary" and "secondary" from a humanities or social sciences perspective.

An Overview of Primary and Secondary Sources in Science and Engineering Disciplines

In the last couple of chapters, we talked about how sources can have qualities that make them scholarly or popular, but now we'll shift to a completely different way of categorizing sources. As you know, in science and engineering, new research discoveries are first documented and shared with other researchers in sources like **journal articles** and **conference papers**. As time passes, information in those sources may get summarized or analyzed and show up in other types of sources such as **review articles**, **newspaper articles**, **websites**, **books**, etc. We have special names for these two phases of information dissemination: **primary sources** describe brand-new, original information, while **secondary sources** provide a summary, synthesis, interpretation, or analysis of pre-existing information.

What Do Science and Engineering Disciplines Consider to be Primary Sources?

A source is considered **primary** if it records information that was "just born"—that is, information that's brand-new, original, and didn't exist in any other form before. In science and engineering disciplines, primary sources usually communicate details about a research project. Typically, the person (or people) who conducted the research project are the authors of the source, which means the source is a **first-hand account** of what the authors researched and learned. So, primary sources contain brand-new research findings documented first-hand by the authors.

A good example of a primary source is a **research article** published in a journal, and that's why research articles are sometimes referred to as **primary articles**, **primary research articles**, or **original articles**. Most **conference papers** are also primary sources. When you think about what you learned about these two sources in Chapters 7 and 9, you'll agree that both describe original, brand-new research that's being shared for the first time. Plus, the authors of these sources are the same people who conducted the research, so they are reporting first-hand on what they discovered. Other primary sources include **dissertations**, **theses**, **data**, and **patents**, because all these sources present brand-new research results written first-hand by the author, creator, or inventor.

For reference, here's a list of sources that are typically considered to be "primary" in science and engineering disciplines:

- research articles
- conference papers
- dissertations
- theses
- data
- patents

What Do Scientists and Engineers Consider to be Secondary Sources?

A source is considered **secondary** if it provides a summary, synthesis, interpretation, or analysis of information gathered from previously existing sources, such as primary sources. So secondary sources do not present newly discovered information. Rather, they summarize and interpret information derived from previously existing sources.

A **review article** is an excellent example of a secondary sources. As you learned in Chapter 8, the authors of a review article gather existing information on a topic—mostly from primary research articles—then write up a review and summary of what was found in those articles. To get a better idea how review articles synthesize pre-existing information rather than describing brand-new information, take look at an excerpt from this review article:

> Zhang, W., et al. (2019). Nanotechnology in remediation of water contaminated by poly-and perfluoroalkyl substances: a review. *Environmental Pollution*, *247*: 266–276.

Here's the excerpt:

> Thanks to their unique structure and physicochemical properties, carbon nanotubes (CNTs) can be potentially applied in energy storage (Kayiran et al.,

> 2004), containment remediation (Colvin, 2003), and drug delivery (Liu et al., 2007). Recently CNTs have been investigated as a potential candidate for the removal of PFASs from water, owing to the intense adsorption affinity between CNTs and organic pollutants (Pan and Xing, 2008).

Notice how the authors are reviewing and summarizing the applications for carbon nanotubes from information they found in *other sources*. They used in-text citations to refer to those other sources and you'd find the corresponding full citations listed at the end of the article.

Two more examples of secondary sources are **reference books** and **textbooks**. Let's take a look at an excerpt from a textbook:

> Lovette, I. J. & Fitzpatrick, J. W. (Eds.) (2016). *Handbook of Bird Biology*. John Wiley & Sons.

Here's how that book defines the traplining behavior that hummingbirds engage in:

> **Traplining**: foraging movements in which a bird rotates through a set of discrete food sources; most often applied to individual hummingbirds that visit a set of flower patches in an established and repeated sequence, allowing the time in between for the flowers to replenish their nectar.

That excerpt provides a simple, concise definition of traplining behavior just like you'd expect in a textbook or a reference book—but all the information in that definition *originally* came from multiple primary research studies conducted from the 1970s to the 1990s establishing traplining as a behavior that hummingbirds engage in. The excerpt doesn't provide citations to all those studies because traplining is now common knowledge and an established fact.

What else is a secondary source? In science and engineering disciplines, **popular sources**—like this *New York Times* article—are considered secondary sources:

> Carey, B. (2022). Can we get better at forgetting? *The New York Times*. https://www.nytimes.com/2019/03/22/health/memory-forgetting-psychology.html

In that news article, the journalist reports on findings from this primary research article:

> Wang, T. H., et al. (2019). More is less: increased processing of unwanted memories facilitates forgetting. *Journal of Neuroscience*, *39*(18): 3551–3560.

So, the journalist took information from a primary source, then wrote up a newspaper article summarizing and simplifying the research for readers of *The New York Times*, thus creating a secondary source.

Or, consider a Wikipedia article, which is another type of popular, secondary source. Wikipedia articles on science and engineering topics offer excellent overviews that have been compiled from both primary and secondary sources. But these articles don't disseminate brand-new, original information.

For reference, here's a list of what's typically considered to be a secondary source in science and engineering disciplines:

- any type of review article
- most scholarly books as well as textbooks, reference books, and popular books
- newspaper articles and other news sources
- magazine articles
- websites
- media (podcasts, TV shows, etc.)

The Relationship between Primary and Secondary Sources

Figure 16.1 shows the relationship between primary and secondary sources in science and engineering, and how newly discovered information that's published in primary sources (such as research articles and conference papers) might get disseminated into secondary sources over time.

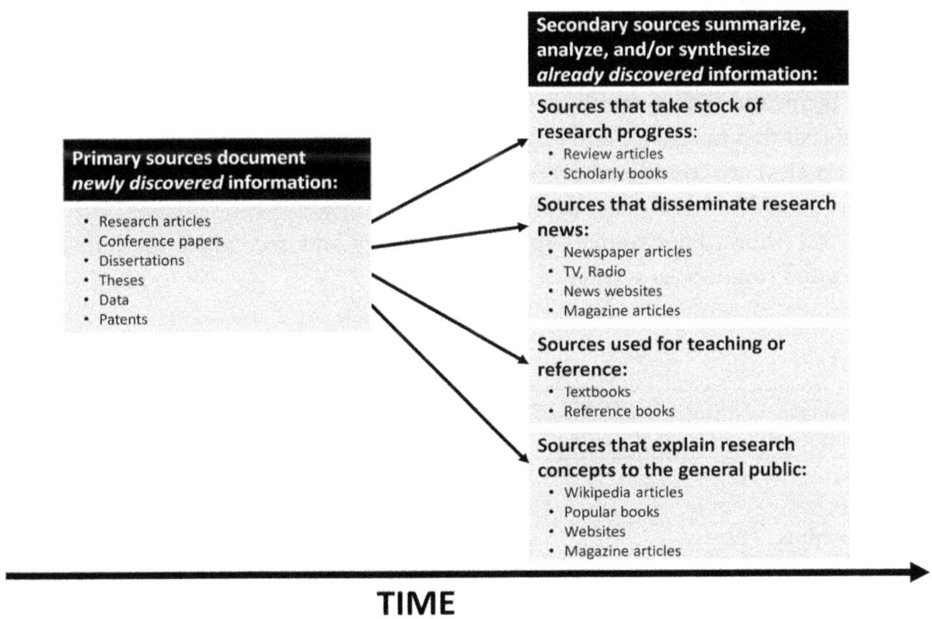

Figure 16.1

Keep in mind there's no guarantee that primary source information *will* end up in secondary sources, and even if it does, it might appear in some secondary sources and not others. So, Figure 16.1 shows various ways information from a primary source *could* get disseminated in secondary sources, not how it *will*.

In general, what makes a source primary or secondary is the nature of the *content* described in the source. If the content describes something that's brand new and unique—and that's never been shared in any other form before—then it's a primary source. If it summarizes and/or analyzes pre-existing information, then it's a secondary source. But, sometimes there's disagreement about what constitutes a primary versus a secondary source. For example, some scientists and engineers would consider certain types of review articles, such as systematic reviews and meta-analyses, to be more like primary sources than secondary sources. And some scientists and engineers consider raw data to be the only type of primary source—everything else is a secondary source.

Primary and Secondary Sources in Other Disciplines

What counts as a primary or secondary source is different in humanities disciplines and some social science disciplines like history and law. Let's discuss these disciplinary differences so you don't end up in a dispute with a historian about what primary and secondary sources are!

In these disciplines, primary sources include letters, diaries, autobiographies, interviews, news articles written at the time of a historical event—as well as original art, literature, photos, films, etc. After all, these types of sources are original and created first-hand by someone. Humanities scholars study these primary sources, analyze and interpret them, and then write about them in *secondary sources* like books and journal articles. Recall that in science and engineering disciplines, journal articles (specifically research articles) are considered primary sources, but in the humanities, most journal articles are considered secondary sources.

Table 16.1 shows how sources split into primary and secondary for humanities disciplines and some social science disciplines.

Table 16.1

Typical examples of PRIMARY SOURCES in the humanities and some social sciences	*Typical examples SECONDARY SOURCES in the humanities and some social sciences*
• Letters • Diaries • Manuscripts • Autobiographies • First-hand accounts, eyewitness accounts, or editorials/opinions on contemporary topics found in newspaper articles, magazines, websites, or other media • Interviews • Literature and works of art such as music, paintings, photography, etc.	• Journal articles • Scholarly books • Any content found in newspaper articles, magazines, websites, or other media that provides an analysis of existing information or a discussion of events that happened in the past

You might have noticed from this table that when you are looking at content from a newspaper, magazine, website, or other media, what counts as a primary source are reports on *current events at the time they happened.* For example, an opinion piece in a newspaper could be either a primary or secondary source, depending on the content and when it was written. Suppose it's early 2020 and you're reading a newspaper opinion piece about how the growing coronavirus pandemic is affecting the economy. That's a primary source because it's someone's opinion about an event that's currently happening. Now suppose it's 2025 and you're reading an opinion piece that looks back over the coronavirus pandemic and how economic issues were handled back in 2020. That's a secondary source because it's providing an analysis and summary of events *after they happened.*

Also, note that just because a source is *about* a humanities or social science topic doesn't mean that the definitions of primary and secondary for those disciplines apply to the source. For example, the journal article shown below is about paintings—a humanities topic:

> Sigaki, H. Y., et al. (2018). History of art paintings through the lens of entropy and complexity. *Proceedings of the National Academy of Sciences, 115*(37): E8585–E8594.

In the humanities, a journal article discussing paintings would normally be considered a secondary source, because the paintings themselves would be the primary source. But in this article, the authors are analyzing data from paintings using scientific methods to arrive at original research findings. Thus, this journal article is a primary source.

You can use the lists of primary and secondary sources provided in this chapter as a guide for what's *typically* considered primary or secondary in different disciplines. But you've also seen that the concepts of "primary" and "secondary" are open to some degree of interpretation. So, if you're working on a class assignment that requires you to find and use primary and/or secondary sources, make sure you know how *your* instructor defines these qualities and ask them to provide examples of sources they consider to be primary or secondary.

Why the Qualities "Primary" and "Secondary" are Useful to Scientists and Engineers

There wouldn't be any point in labeling a source as "primary" or "secondary" unless those qualities were useful in some way—and they are. One reason primary sources are particularly important in science and engineering is that they contain original findings written first-hand. Perhaps you've heard the advice "Go to the original source." That means you should look at information in its original form to read and interpret it for yourself, rather than relying on an interpretation you found in a secondary source. Keep in mind that any time information is summarized, analyzed, or interpreted—as it is in secondary sources—there's a chance it will be oversimplified and/or misrepresented. After all, the further information gets from the original source, the more it can stray from its original meaning or intent.

In fact, straying from original meaning and intent is a fairly common problem in science journalism, as you see in this example:

> In 2018, psychology PhD student William McAuliffe co-published a paper in the prestigious journal *Nature Human Behavior.* The study's conclusion—that people become less generous over time when they make decisions in an environment where they don't know or interact with other people—was fairly nuanced.
>
> But the university's press department, perhaps in an attempt to make the study more attractive to news outlets, amped up the finding. The headline of the press release heralding the publication of the study read "Is big-city living eroding our nice instinct?"
>
> From there, the study took on a new life as stories in the press appeared with headlines like "City life makes humans less kind to strangers."

This example was taken from:

> Resnick, B. (2019). Hyped-up science erodes trust. Here's how researchers can fight back. *Vox.* https://www.vox.com/science-and-health/2019/6/11/18652225/hype-science-press-releases.

The author goes on to point out:

> Frequently, stories about scientific research declare an exciting new treatment works but fail to mention the intervention was performed on mice. Or stories that breathlessly report what the latest study finds on the health benefits (or risks) of coffee, without assessing the weight of the available evidence.

So, if you're reading a secondary source and suspect that the author's interpretation is sensationalized, oversimplified, or just wrong, then you can track down the primary source(s) the author referred to and read them for yourself to decide on their meaning and significance. Primary sources allow us to verify and interpret original content for ourselves, which is particularly critical in situations where we feel that the author of a secondary source may have portrayed original content inaccurately.

Secondary sources—such as review articles and textbooks—are useful for different reasons. These sources provide an overview of what we know about a topic, so they tend to be easier to read and understand. In fact, they're ideal for learning about a topic that's new to you. Remember the *Handbook of Bird Biology* discussed earlier in this chapter? You saw that it provides a concise definition of hummingbird traplining behavior that's easy for students to read and comprehend. It would take a lot more time and effort for students to learn about traplining if they had to find and read all the primary research articles where traplining was first researched and reported on.

Secondary sources are also perfect for when you need a broad overview of a topic. For example, if you need some brief, general information about what cryptography is, a Wikipedia article on the topic will serve you much better than trying to wade through a bunch of primary research articles about cryptography.

Just know that the reasons why scientists and engineers value primary and secondary sources isn't exactly the same for humanities scholars, historians, etc. As you've learned, those scholars categorize and use primary and secondary sources a bit differently.

To sum up, what should science and engineering students know about primary and secondary sources?

1. A source is considered primary if it records information that was "just born"—that is, information that's brand new, original, and didn't exist in any other form before.
2. A secondary source takes information from existing sources and summarizes, interprets, synthesizes, or analyzes that information.
3. The types of sources considered to be primary or secondary in science and engineering disciplines are not the same as in other disciplines (such as the humanities and some social sciences).
4. Primary and secondary sources have different qualities and uses.

What's Next?

In the next chapter, you'll learn that when a source is considered **credible** that means there's reason to believe it's accurate and reliable. You'll also learn how to identify attributes in a source that signify credibility.

CHAPTER 17
CREDIBLE SOURCES

The quality of being "credible" is a slippery concept to grasp, but this chapter will fix that. You'll find out what attributes signify a source is credible, and you'll learn that assessing credibility involves gathering evidence as well as using your intuition and experience. You'll also learn that establishing the credibility of a source is not a guarantee the source will contain accurate, reliable information—instead, it's more like placing a calculated bet. Also, the way you judge the credibility of sources will evolve over time as you move from being a student to an expert researcher.

What Does It Mean for a Source to Be "Credible"?

A source is considered **credible** if it's reasonable to assume it will provide accurate and reliable information. But how do you know when that's reasonable to assume? You'll need to examine the source for certain qualities and attributes to make that determination. For example, you might be able to establish that the source is scholarly, or that it was written by someone with appropriate expertise. Or you might note whether the source provides lots of explanations, examples, and evidence, because that can indicate credibility. You'll learn about these attributes—and more—shortly.

Being "Credible" Doesn't Guarantee Accuracy, but It's Still a Useful Concept

You just learned that credible sources have attributes that make us confident they contain trustworthy information. Suppose you're examining a source and you decide it's reasonable to assume it contains accurate and reliable information—in other words, you determine it's credible. Unfortunately, that's no guarantee the source *does* contain accurate and reliable information. So, you may be wondering: "If being 'credible' is no guarantee that a source is accurate and reliable, then why bother discussing this quality at all? Wouldn't it be better to make *certain* that a source contains reliable information?" It *would* be preferable to be certain, but most of the time that's not possible. To see why, let's consider this source, a research article:

> Manikkam, M., et al. (2013). Plastics derived endocrine disruptors (BPA, DEHP and DBP) induce epigenetic transgenerational inheritance of obesity, reproductive disease and sperm epimutations. *PLoS One*, *8*(1): e55387.

To be certain the information in that source is accurate and reliable, you'd need to be an expert in the field of transgenerational inheritance of health conditions caused by exposure to endocrine disruptors. Then you'd be able to fully understand this research article and assess whether the research project it describes was well-designed, and whether the authors collected enough relevant data and analyzed those data properly to arrive at sound conclusions. To be even more certain of this article's reliability, you might also want to check the quality of the sources cited and how the authors used information from those sources to support their research findings. It wouldn't be a bad idea to consult other expert scholars as well, to see if you all agree that the article is accurate and reliable.

If you were already an expert on the subject matter of this article, then you *could* potentially take the time to thoroughly assess it and determine whether it's sound. In fact, when experts use sources within their narrow area of expertise, it's common for them to carefully assess the rigor of those sources.

The problem is, we need to consult sources *outside* our areas of expertise all the time. For example, we use sources to solve a variety of problems at work as well as problems in our personal lives—such as making choices about medical care, investing money, deciding who to vote for, deciding what products to buy to get the best value for our money, etc. Our skill at selecting and consulting high-quality sources impacts our career success, health, and well-being. However, most of the time, it's completely unrealistic to assess those sources in the way described above. It's too time-consuming—plus it would require everyone to become an expert on everything (which is obviously impossible). All we can do, then, is examine sources for attributes that are *associated* with rigor in order to gauge which sources are a safe bet to rely on. So, if you examine the attributes of a source and conclude that it's credible, it's like you're placing a calculated bet that it'll contain accurate and reliable information.

Furthermore, we instinctively spend more time assessing credibility when the stakes are high. For example, suppose you injure your foot. As you don't have the appropriate expertise to diagnose and fix your injury, you must rely on an information source that does, namely: a doctor. So, you go to a doctor who says you'll need intricate surgery to fix your foot, followed by many months of physical therapy. The doctor also explains that the surgery has risks. Sometimes the surgery isn't effective, plus there've been a few cases where patients had worse foot problems *after* the surgery. Because your mobility is a high-stakes issue, you'd want to spend more time and effort assessing the credibility of the advice this doctor gave you. Specifically, you might:

1. Get a second or third opinion from other doctors, and specifically target doctors who specialize in foot injuries *and* who have a lot of experience.
2. Find and read journal articles about the surgical procedure and whether there are alternative, less-risky therapies.
3. Zero in on journal articles that are **systematic reviews** or **meta-analyses**, as those articles have qualities that make them highly credible (see Chapter 8 for more information).

While you'll never have the necessary expertise *to know for sure* the best course of action for fixing your foot, you can gather information, assess its credibility, and then make a calculated bet on what's best for you. And, you'll spend *more* time assessing credibility when the stakes are high, like they are in the example just described.

There will, of course, be times when you judge an information source to be credible and then find out it contains flawed information. For example, the article below (which we discussed in Chapter 14) has lots of attributes indicating it's a credible source, yet it turned out to be flawed, and the publisher retracted it:

> AlShebli, B., et al. (2020). The association between early career informal mentorship in academic collaborations and junior author performance. *Nature Communications*, 11(1): 5855.

So, making calculated bets isn't foolproof—that's why they're called *bets*! Nevertheless, taking the time to assess the credibility of a source and make a calculated bet on its accuracy and reliability will pay off most of the time, and it's certainly better than using any source at random. That's why it's important to be skilled at identifying credible sources.

Several attempts have been made to create simple checklists for students to follow to help them establish whether a particular source is credible. While these checklists can be useful, none work perfectly in all situations, nor do they capture how remarkably complex and nuanced judging credibility really is. To get an idea of the different variables that might come into play when making credibility judgments, take a look at Table 17.1.

Table 17.1

How we might judge the source's *author*	• Personal reputation, character • Expertise, position to know • Educational credentials • Years of experience • Prestige, authority • Motivation, intent • Composure, neutrality • Repercussions for being wrong or deceptive
How we might judge the source's *producer* (meaning the publisher, organization, agency, etc.)	• Reputation • Longevity, stability • Prestige, authority • Motivation, intent • Transparency (clear policies, disclosures, contact info, physical address, etc.) • Gatekeeping and quality control activities (editorial review, peer-review, fact-checking, etc.) • Size of audience producer is accountable to • Repercussions for being wrong or deceptive

Continued

Table 17.1 *Continued*

How we might judge the source's *container* (that is, whether it's a book, magazine, website, etc.)	• Scholarly or popular? • Primary or secondary? • Hurdles to publication (peer-review, editorial review, printing and distribution costs, etc.) • Appearance (professionalism, seriousness, neatness, layout) • Presence or absence of advertising
How we might judge the source's *content*	• Utility • Accuracy • Informativeness • Aesthetics, appearance • Organization, navigation, structure • Objectivity, neutrality, balance, fairness • Currency • Coherence, clarity • Corroboration of content in other sources • Writing style (professional, technical, etc.) • Coverage, completeness, depth • Presence of evidence and examples • Motivation, intent • Methodological rigor • Transparency, cited references • Popularity • Reviews, awards, or endorsements • Size of audience the source is accountable to • Type of audience the source is accountable to • Number of editions
Dispositions the *reader* brings to judging credibility	• Familiarity with the individual source • Familiarity with the source type (container) • Familiarity with the source authors • Familiarity with the source publisher/producer • Amount of experience judging credibility • Whether source content supports reader's existing views • Reader's interest in content • Reader's cultural background

Some of the evidence in this table was gathered from: Metzger, M., et al. (2003). Credibility for the 21st century: integrating perspectives on source, message, and media credibility in the contemporary media environment. *Annals of the International Communication Association, 27*(1): 293–335. DOI: 10.1080/23808985.2003.11679029

So, the evidence you consider will depend on your previous knowledge and experience with the author, publisher, and source container, as well as what you think of the actual content. Your own disposition towards the material will also factor in to how you judge it. In fact, there's so much evidence that *could* come into play, we can't cover all of it, at least not in this chapter. So, we'll start by focusing on strategies and techniques that are helpful to *novice researchers*. Later in this chapter we'll explore how those strategies might change as you gain more experience and subject expertise.

To Judge a Source's Credibility, First Establish if It's Scholarly

For novices, the best way to start judging credibility is to ask yourself: "Is the source scholarly?" That is, is it a journal article, conference paper, or a scholarly book? If so, then you can assume the source is credible. Recall in Chapter 14 you learned about the various qualities scholarly sources have—for example, they're written by expert scholars, those scholars provide support for their claims, and their manuscripts undergo editorial review and/or peer review before being published. Those qualities make it *reasonable to assume* scholarly sources are credible, in spite of the fact that there are instances when those sources are imperfect and contain flawed information. In fact, you can think of scholarly sources as a type or subset of credible sources, like you see in Figure 17.1.

Keep in mind if the source you're examining is a journal article, you should make sure it's not from a predatory journal. Articles from predatory journals are not scholarly, nor are they credible. See Chapter 6 for information about how to identify predatory journals.

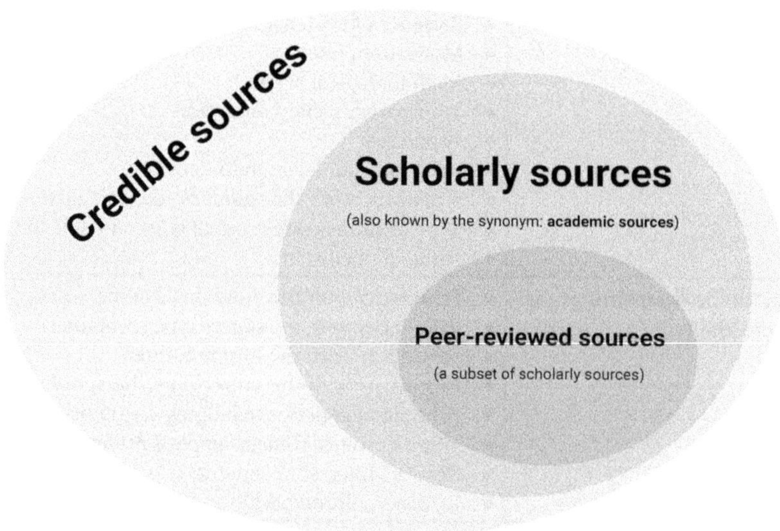

Figure 17.1

If a Source *Isn't* Scholarly, then Examine It for Attributes that Signify Credibility

If you're evaluating a source that you know is *not* scholarly, or if you can't tell whether the source is scholarly, then you'll need to examine it for attributes that signify credibility. Here are some attributes that might factor into your judgment:

- Does the author have appropriate expertise and/or are they in a position to know what they are talking about?

- Is it in the author's interest to protect their reputation as an accurate, reliable writer? Is it likely they'd be held accountable for presenting inaccurate information?
- Does the source's content stand up to background checking and/or corroboration?
- Does the author provide logical, rational arguments supported by evidence and examples—perhaps even citing outside sources to bolster their claims?
- Is the author's writing style neutral, objective, coherent, and forthright?

Let's discuss these attributes in more detail, so you can get a good feel for them.

Does the Author Have Appropriate Expertise and/or Are They in a Position to Know What They Are Talking About?

For a source to be credible, the answer to this question should be "yes"—but how do you make that determination? Suppose you're examining the book below for credible qualities, including whether the author is an expert on the topic they're writing about. To do that, conduct a Google search on the author to see if you can find out more about them.

> Carey, N. (2012). *The Epigenetics Revolution: How Modern Biology is Rewriting Our Understanding of Genetics, Disease, and Inheritance*. Columbia University Press.

When you search Google for the author of this book, Nessa Carey, you'll find she has a PhD degree in virology, she was a senior lecturer in molecular biology at Imperial College London, and she was a panelist at the 2019 Nobel Week Dialogue. In other words, she seems to have relevant expertise, so it would be reasonable to assume this book contains accurate and reliable information—or in other words, it's a credible source.

Or, consider whether someone is in a good position to know about a topic. Let's take information from this *New York Times* article as an example:

> Imbler, S. (2021). When an eel climbs a ramp to eat squid from a clamp, that's a moray. *The New York Times*. https://www.nytimes.com/2021/06/22/science/moray-eels-eat-land.html

The journalist who wrote this news article is reporting on a research study about how moray eels can hunt and eat prey out of the water. In the study, squid were used as bait to entice the eels out of the water. Within the news article the journalist also reports the following:

> And morays climbing out of water came as no surprise to some observers. Lana Sinapayen, an artificial life researcher who grew up in the Caribbean island of Martinique, said local fishermen often caught morays by placing squids on the shore and waiting for the eels to arrive. "You only need a solid stick to take your pick," she wrote in an email. Dr. Sinapayen was not involved in the research but wanted to emphasize that many local people have long known that morays can hunt on land.

Now let's talk about the credibility of the fishermen described in the excerpt above. Are those fishermen a credible source of information on eel feeding behavior? Yes! The fishermen have direct, first-hand experience enticing eels onto land with bait. This traditional ecological knowledge carries a great deal of weight—in fact, it's enough to consider this information about eels to be credible.

Is It in the Author's Interest to Protect Their Reputation as an Accurate, Reliable Writer? And, Is It Likely They'd Be Held Accountable for Presenting Inaccurate Information?

Consider whether the author of a source needs to foster and protect a good reputation and/or whether they'd suffer negative repercussions for presenting bad information. For example, suppose you're a software engineer who posts answers to the online coding forum, Stack Overflow, and your posts regularly get lots of positive votes. In Stack Overflow, positive votes get you reputation points, and reputation points get you certain privileges:

> Reputation is a rough measurement of how much the community trusts you; it is earned by convincing your peers that you know what you are talking about. The more reputation you earn, the more privileges you gain and the more tools you have access to on the site—at the highest privilege levels, you'll have access to many of the same tools available to the site moderators. That is intentional. We don't run this site; the community does![1]

It would be in the interests of anyone with a good reputation on this site to try to maintain that reputation and the privileges it confers, rather than having those privileges revoked and losing their standing. Plus, users with a good reputation are probably proud of their reputation and enjoy helping others, otherwise they wouldn't have bothered to build up reputation points, so this makes them credible!

To give another example, when scientists and engineers working at colleges or universities publish their research findings, it's in their interest to be accurate and reliable. If they regularly publish erroneous information, or if they intentionally fabricate their results, it's likely they'll get fired. Here's a headline from the Retraction Watch website showing just that:

> Oransky, I. (2013). *Leiden University fires employee for research fraud, two retractions to follow.* Retraction Watch. https://retractionwatch.com/2013/08/15/leiden-university-fires-employee-for-research-fraud-two-retractions-to-follow/

---

[1]From: What is reputation? How do I earn (and lose) it?—Help Center. Stack Overflow. (n.d.). https://stackoverflow.com/help/whats-reputation.

This headline makes it clear that research fraud *does* happen; however, the repercussions are dire if you get caught, so researchers are better off maintaining a reputation for being credible.

Note that for some sources, there's no particular person named as an author, but rather an organization or agency is taking responsibility for the information, for example:

> U.S. Department of the Interior. (2021). *Invasive Species Strategic Plan*. https://www.doi.gov/sites/doi.gov/files/doi-invasive-species-strategic-plan-2021-2025-508.pdf

For this publication, we can consider the U.S. Department of the Interior to be the author, and we can consider it to be credible. Why? Because U.S. government agencies have a reputation to uphold as credible sources of information. If they provide misinformation, it could cause a scandal—especially because they're funded by taxpayer dollars.

Basically, if it seems that it's in an author's interest to be an accurate and reliable source of information, that counts as evidence their works are credible. It doesn't matter if the author is a person, a group of people, an organization, an agency, or some other entity.

Conversely, you should be especially wary of information that can't be traced to a particular person, organization, agency, or any other entity. When information comes from a completely unknown, untraceable author, there's no incentive for that author to strive for accuracy. In other words, there's no accountability. After all, if the author provides misinformation, nobody can call them out on it because nobody knows who they are.

This was a problem for the website Yahoo Answers, where anyone could ask a question and anyone could provide an answer. As the people answering questions didn't need to provide their real names or show any credentials, there was no accountability and often the answers were terrible. In fact, this lack of accountability became an intractable problem for Yahoo Answers and eventually it shut down on May 4, 2021, as you see in the headline below:

> Statt, N. & Peters, J. (2021). *Yahoo answers will be shut down forever on May 4th: the end of a long and storied internet era (of bad questions and even worse answers)*. The Verge. https://www.theverge.com/2021/4/5/22368488/yahoo-answers-shutdown-may-4-internet-era-over-rip

So now you may be wondering: "What about Wikipedia?" Wikipedia articles don't provide named authors, which means the authors needn't worry about protecting their reputation as accurate, reliable writers. Except that Wikipedia has strategically put other measures in place to make these anonymous authors feel accountable for their contributions. For example, claims and facts must be backed up with cited references, and if this is not done properly, the content is likely to be removed or edited. This policy motivates contributors to do good work—otherwise they're wasting their time writing content that will get taken down. There are also edit tracking systems and article grading systems in place. Plus, Wikipedia has a hierarchy of trusted, volunteer administrators and "bureaucrats" who perform various functions to monitor the site, including the ability to block contributors who are abusing the system. All these measures give Wikipedia articles much more credibility than they'd have otherwise, and certainly more credibility than Yahoo Answers. While you should always be careful about the accuracy of content in Wikipedia articles, overall it's a phenomenally useful source given that it's authored by thousands of decentralized, anonymous contributors.

Does the Source's Content Stand Up to Background Checking and/or Corroboration?

Background checking can take a variety of forms. One way to conduct a background check is to investigate what's being said about a source. To see how that works, let's investigate the online news website, Mongabay (https://news.mongabay.com/), which specializes in environmental science and conservation news. The easiest and best way to check up on a source like this is to look for a Wikipedia entry for it and read what it says. When you read the Wikipedia entry for Mongabay, you'll learn it's gotten several awards, plus it's funded by well-known philanthropic foundations, and it's rated highly as a non-profit. This evidence suggests it's a credible source.

Background checking can also take the form of looking for reviews. Suppose you're interested in this book, but you're not sure if it's credible:

> Winchester, S. (2018). *The Perfectionists: How Precision Engineers Created the Modern World.* Harper.

You could, of course, do some research on the author to see whether they have relevant expertise, like you learned to do earlier in this chapter. But you could also look for reviews of the book to see what's being said about it. To find book reviews, conduct a Google search for the book along with the word "reviews" like this:

- **"The perfectionists" Simon Winchester reviews**

If you find reviews, then you can read them and learn whether the reviewers thought the book was informative and well-researched—and thus credible.

Going to the original source—which we discussed in the last chapter—is another form of background checking. Suppose you're writing a research paper that explores whether the use of social media is contributing to teenagers' levels of anxiety and depression. In the course of your research, you come across this newspaper article:

> Hughes, V. (2022). Does social media make teens unhappy? It may depend on their age. *The New York Times.* https://www.nytimes.com/2022/03/28/science/social-media-teens-mental-health.html.

The information in that article would be useful to cite in your research paper, and it's coming from *The New York Times*, which is generally considered to be a credible news source. However, you notice the news article is reporting on findings from a scholarly journal article, and it even provides a direct link to that journal article. You can follow the link to the journal article (which is the original source) and read it for yourself to see what the researchers' found—and confirm whether the *New York Times* reporter accurately represented the findings of the research article:

> Orben, A., et al. (2022). Windows of developmental sensitivity to social media. *Nature Communications*, *13*(1): 1–10.

Of course, if you've gone to the trouble to track down and read the original source, then you should cite the original source instead of *The New York Times* article. In science and engineering disciplines, citing the original, primary source (rather than citing a popular, secondary source like a newspaper article) is preferable. Why? Because then you're giving credit to the *original creators* of the information. Plus, you can present your *own* interpretation of information from the original source, rather than presenting your interpretation of the *New York Times* reporter's interpretation.

You can also seek out corroborating information to help you judge whether a source is credible. Suppose a friend told you that all bats are blind, and you weren't sure this was true. You could check other sources to see if they provide corroborating evidence. A quick Google search reveals lots of sources saying that bats are *not* blind. It seems, then, that your friend is *not* a credible source of information on bats.

Also, at the beginning of this chapter, you saw an example of gathering corroborating information when we discussed actions you might take if you needed risky foot surgery. For example, you might get a second or third opinion from different doctors and/or conduct your own research for journal articles about the surgery or alternative treatments.

Or, suppose you're reading about a news event, but suspicious of its accuracy. You could look up the news event in other news sources to see if it's described basically the same way in each. If so, that's an indication the news is credible. You can also double-check the veracity of news using special fact-checking websites, such as FactCheck.org, Politifact.com, Snopes.com, etc. To find these sites, just do a Google search on **fact-checking websites**.

Does the Author Provide Logical, Rational Arguments Supported by Evidence and Examples—Perhaps Even Citing Outside Sources to Bolster Their Claims?

Whenever an author presents readers with strong arguments that are rational and are backed up with explanations, examples, and evidence, this enhances their credibility. Let's look at content in this publication as an example:

Tam, C. K., et al. (2021). *US Geological Survey Invasive Species Research—Improving Detection, Awareness, Decision Support, and Control* (No. 1485). US Geological Survey.

Specifically, let's look at the table of contents and a snippet from the section on invasive aquatic plants (Figure 17.2). Notice from the table of contents these researchers thoroughly cover different types of invasive species as well as methods for detecting them. And, within the chapter on invasive aquatic plants, the authors discuss specific examples. In other words, these authors are putting forward information that's *substantiated.*

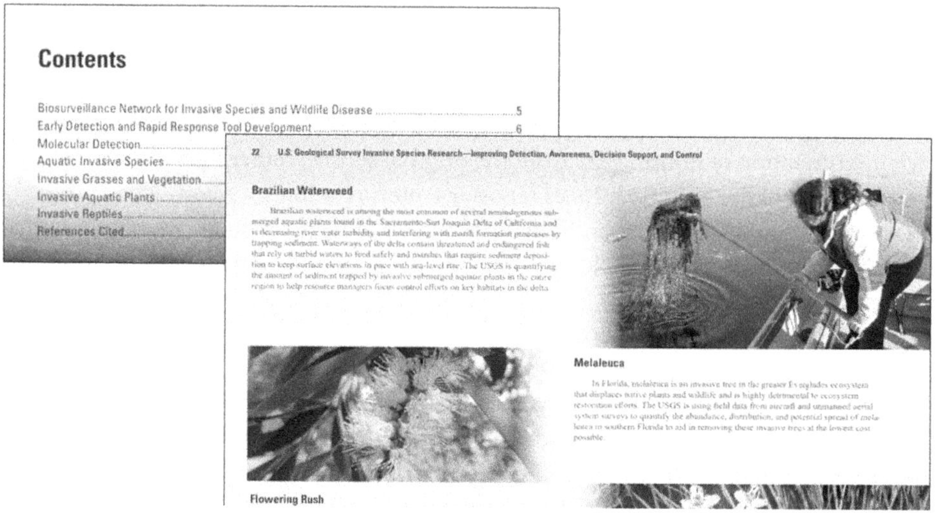

Figure 17.2

In fact, one of the reasons why research articles and conference papers are considered credible sources is because the authors of those sources provide so much substantiation. The authors describe the methods they used, the evidence they gathered, and how they analyzed that evidence to reach conclusions. Providing those details allows for the study to be *reproduced* if there are questions about the validity of its findings. The United States National Science Foundation (NSF) defines **reproducibility** as "the ability of a researcher to duplicate the results of a prior study using the same materials as were used by the original investigator." Furthermore, NSF considers reproducibility to be "a minimum necessary condition for a finding to be believable and informative." In other words, when authors don't provide a lot of detail and explanation, their study can't be reproduced, which makes it less believable, and thus less credible. On the other hand, when you provide a lot of detail and explanation, you're showing your intent to be transparent and honest. Plus, you're indicating that you're so confident in the strength of your work that you're happy to share all the details and have those details scrutinized. So, providing a lot of detail and explanation sends a strong message about the credibility of a source.

Referring to outside sources and citing those sources is another way to support claims and augment evidence. When an author cites outside sources in their writing, it's as if the author is saying: "I'm so confident in the accuracy of the information I'm writing, that I invite you to double-check my sources if you don't believe me." So, including citations to source material is another practice that enhances credibility.

Of course, an author who is deliberately trying to misinform or manipulate might do so by presenting arguments that appear sound and by offering evidence and citing sources, it's just that they might present *bad* evidence and cite *bad* sources! So, be aware that non-credible sources may try to imitate credible sources in this way.

Is the Author's Writing Style Neutral, Objective, Coherent, and Forthright?

The purpose of most credible sources is to inform, using language that's neutral and objective, and that's exactly the tone you'd expect from an author whose intention is to be honest and forthright.

Let's look at an example from this article in the magazine *Scientific American*:

> Blaustein, A. (2022). Rechargeable molten salt battery freezes energy in place for long-term storage. *Scientific American*. https://www.scientificamerican.com/article/rechargeable-molten-salt-battery-freezes-energy-in-place-for-long-term-storage/

First, notice how the title of the article is matter-of-fact and neutral. Now, take a look at this excerpt:

> Most conventional batteries store energy as chemical reactions waiting to happen. When the battery is connected to an external circuit, electrons travel from one side of the battery to the other through that circuit, generating electricity. To compensate for the change, charged particles called ions move through the fluid, paste or solid material that separates the two sides of the battery. But even when the battery is not in use, the ions gradually diffuse across this material, which is

> called the electrolyte. As that happens over weeks or months, the battery loses energy. Some rechargeable batteries can lose almost a third of their stored charge in a single month.

There's nothing in this excerpt that suggests anything other than that the writer's intent is to inform the reader in a way that's neutral and objective. Something else to note about this excerpt is that it's well written and coherent, and that's also a sign of credibility! But be aware that an author who is trying to make bad information *appear* credible may mimic a credible writing style by using a neutral, objective, and forthright tone. So, you can't rely on this attribute on its own to establish credibility.

When It Comes to Credibility, What Should You Be Especially Wary of?

You should be especially wary of the credibility of sources that discuss controversial, political, or highly emotional topics. Obviously, there are credible sources that discuss these kinds of topics, but non-credible sources will *regularly and actively* exploit highly charged topics for motives other than to inform—such as to persuade you to take a side, take some sort of action, or buy something.

For example, consider a website that contains content about how to "hack your metabolism" to lose weight. That content is emotionally charged for people who want to lose weight, and the website author can take advantage of that. For example, the author can write in a way that will play upon the reader's insecurities and manipulate them into buying expensive weight-loss products.

In fact, the main motive for many websites is to get you to visit that website, spend time on it, and expose yourself to advertising. This is how **clickbait** works. It presents you with a headline or an image that causes you to experience surprise, fear, insecurity, outrage, sadness, or disgust, so you feel compelled to click on it. For an example, let's revisit a headline you saw back in Chapter 4:

> How using Facebook could increase your risk of cancer. (2009). *Daily Mail.* https://www.dailymail.co.uk/health/article-1149207/How-using-Facebook-raise-risk-cancer.html

That's a sensationalized headline and lots of people probably clicked on it to read what the article had to say. This resulted in increased traffic to the publisher's website, allowing them to attract advertisers and thus revenue. So, be *more* careful assessing credibility when the source covers a highly charged topic, and be especially on guard if the topic is discussed in a way that's inflammatory or sensationalized.

Another thing to be wary of is writing that seems deliberately obtuse, jargon-laden, and verbose. Authors who use lots of flowery language and can't write a clear sentence often do so to cover for the fact they have little of consequence to say. This writing style intimidates readers into thinking they just aren't smart enough to understand the brilliance of the author, instead of reaching the more appropriate conclusion that the author is a bad writer who is dressing up trivial information to sound important. If you are reading a source and it seems like the author has made no effort to communicate their ideas clearly, plus they are using lots of big words, or coining lots of their own terminology, then be on your guard.

When You Search Google, Is the Top Result the Most Credible?

Google doesn't know what's credible and what's not, it just uses algorithms to guess at which results will be the most useful, and it ranks those first. So, the first result might be the most credible, but it just as easily might not. Novice researchers tend to put a lot of faith in Google delivering the best, most credible results first, while expert researchers have scrapped that expectation long ago. Instead, expert researchers scan over their Google results to make their *own* choices about what to select. How do they choose? By gleaning information from the titles, URLs, and snippets of their results. After a quick scan, they click on the result that looks most suitable for their needs.

Let's Review How to Establish Credibility and Discuss What to Do When You're Uncertain

Here's how you should approach assessing a source's credibility:

1. Try to establish if the source is scholarly; if so, then you can assume the source is credible.
2. If the source isn't scholarly, or if you can't tell if it's scholarly or not, then look for attributes that signify credibility:
    a. Does the author have appropriate expertise and/or are they in a position to know what they are talking about?
    b. Is it in the author's interest to protect their reputation as an accurate, reliable writer? Is it likely they'd be held accountable for presenting inaccurate information?
    c. Does the source's content stand up to background checking and/or corroboration?
    d. Does the author provide logical, rational arguments supported by evidence and examples—perhaps even citing outside sources to bolster their claims?
    e. Is the author's writing style neutral, objective, coherent, and forthright?
3. Be especially wary of sources that discuss controversial, political, or highly emotional topics—especially if those topics are discussed in a way that's inflammatory or sensationalized, or that cause you to experience surprise, fear, insecurity, outrage, sadness, or disgust.
4. When you search Google, never assume that it's presenting you with the most credible, relevant results first.

Sounds fairly easy, right? The problem is when you're examining non-scholarly sources, the process of looking for attributes that signify credibility can be pretty tricky. It would be great if those attributes were easy to investigate, clear-cut, and we could just check off whether a source had them all or not, and this would determine whether the source was credible. But those attributes don't constitute a checklist. As you've seen, they're not all weighted equally, nor does a source necessarily need each of those attributes to be considered credible. Rather, you have to assess sources on a case-by-case basis, gauging attributes in order to make a calculated bet.

So, you might be wondering, "How will I know when I've found enough attributes to declare a source credible?" The fact is that establishing credibility can be an ambiguous process. You might examine a source and come away feeling quite confident it's credible, or you might feel only somewhat confident, or you might be completely flummoxed about whether the source is credible or not. Getting comfortable with some ambiguity is an important life skill, and judging credibility gives you practice with this life skill—so embrace it! When you spend time examining sources and making calculated bets about whether they're credible (or not), the more you'll get a feel for the process, and the better you'll get at it. Don't avoid it just because you don't always come away with a confident assessment. Instead, jump in and assess sources as best you can. And keep in mind that whenever you're wrong, you'll learn something from the experience that will help you get better at assessing credibility. Also, don't forget that if a particular source doesn't offer clear evidence of being credible, then you can always just ditch it and look for another source that's more clear-cut.

Also consider that when you're in doubt about the credibility of a source, this offers a learning opportunity; ask someone with more experience (such as your instructor or a librarian) what *they* think about the source and find out how they'd judge it. When seeking advice, make sure the person you're consulting is aware you already tried to judge the source's credibility on your own. Explain to them how you looked for evidence but came up short. Then, make sure they explain to you why *they* think the source is or isn't credible. That way you will learn from each other! This is important because experienced researchers tend to forget how difficult it is for novice researchers to judge credibility—nor do they understand where novice researchers are getting stuck. If we don't have open conversations about this issue, we'll never unravel why establishing credibility is such a difficult skill to learn—or to teach!

Last but not least, if you look at a source and you can't immediately detect that it's scholarly or that it has other credible attributes, and you don't have time to investigate it, then don't assume it contains reliable information. Instead, withhold judgment. Be skeptical until you've gathered more information and can make a calculated bet on its credibility.

The Way You Judge Credibility Will Change over Time

Earlier in this chapter you learned that if you're able to determine that a source is scholarly, then you've confirmed it's credible. When you're a novice, you may need some time to examine a source to determine if it's scholarly and thus credible. But expert researchers can usually tell in an instant if they're looking at a scholarly source like a journal article, conference paper, or scholarly book. And, if they're in any doubt, they just look to see if the source was published by a reputable, scholarly publisher. But how do they know which publishers are reputable? Over time, researchers start to recognize and remember the names of publishers who are known for consistently publishing high-quality, reliable sources. Often these publishers specialize in publishing scholarly content, which involves managing a diligent editorial process to ensure the content they publish is high quality (see Chapter 14 for more information about scholarly editorial processes such as peer review). So, the publisher is adding value to the content they publish by taking measures to ensure it is high quality, and as they do this, they build up their brand recognition as a producer of credible sources. As you learn and remember these publishers, you gain a quick way to judge credibility! For example, remember this book we discussed earlier in this chapter?

> Carey, N. (2012). *The Epigenetics Revolution: How Modern Biology is Rewriting Our Understanding of Genetics, Disease, and Inheritance*. Columbia University Press.

An expert researcher would immediately know that book is credible simply because it's published by a reputable scholarly publisher, Columbia University Press. They wouldn't need to investigate the expertise of the author, or look for glowing reviews of the book, or investigate other credible attributes of the book. They'd know the book was credible just from the name of the publisher. Over time you'll start to remember more publishers, organizations, societies, agencies, and institutions that have built a reputation for producing credible sources, and you'll be able to use this shortcut for judging credibility.

In the meantime, you can use your background-checking skills to investigate a particular publisher. For example, if you looked up Columbia University Press in Wikipedia (Wikipedia.org/wiki/Columbia_University_Press), you'd learn that:

> Columbia University Press is a university press based in New York City, and affiliated with Columbia University. It is currently directed by Jennifer Crewe (2014–present) and publishes titles in the humanities and sciences, including the fields of literary and cultural studies, history, social work, sociology, religion, film, and international studies. Founded in 1893,[2] Columbia University Press is notable for publishing reference works, such as *The Columbia Encyclopedia* (1935–present), *The Columbia Granger's Index to Poetry* (online as *The Columbia World of Poetry Online*) and *The Columbia Gazetteer of the World* (also online) and for publishing music.[3]

So, this tells us Columbia University Press has been around a long time (since 1893) and they publish scholarly (and thus credible) sources.

Over time, the way you judge credibility will evolve in other ways as well. When you're a student and conducting research for the first time, most sources are unfamiliar to you. But as time passes and you become immersed in your career, you may find you use a lot of the same journals, books, news sources, and websites over and over again. You'll get familiar with those sources and have a good feel for their credibility. It might be fairly rare that you need to seek information from sources you know little about—so you won't need to spend as much time and effort assessing credibility.

In fact, as you gain expertise, your focus will shift to assessing the rigor and utility of the *content* in the sources you use. For example, suppose a clinical researcher is trying to determine whether psilocybin offers an effective treatment for patients suffering from post-traumatic stress disorder. That researcher might look for journal articles on this topic to find an answer. Let's say that one of the articles the researcher found describes a double-blind, controlled, clinical trial with 400 participants, while another article describes an exploratory study with ten participants. Both of these journal articles are scholarly sources and thus credible, but of the two, the conclusions of the first article will carry more weight because of the robust study design and the larger sample size of participants. So, an experienced clinical researcher would focus on assessing the utility and rigor of the *actual content* in different sources.

But as a student, you need to start somewhere, and the best place to begin is to get better at identifying scholarly sources and acquiring a good feel for attributes that signify credibility. This will allow you to make a calculated bet that a source is (or isn't) accurate and reliable. As you get more practice with this process, you'll get better at it and need to devote less time and mental energy to it. In other words, your ability to discern credible attributes in a source will shift from being a conscious, deliberate process to a much more intuitive process. Then, as you gain more experience and subject expertise, you'll be able to shift your mental energy to assessing the rigor and utility of the *content* delivered in a source, and whether that content is suitable for your needs.

Should You *Only* Use Credible Sources?

Whether you should seek and use credible sources really depends on your information need. You'll probably need to use credible sources for most of your coursework, and you'll definitely need to use credible sources throughout your career, and for resolving various other issues in your life.

However, for lots of low-stakes information needs, we might not spend much time assessing credibility, or worrying much about credibility. For example, suppose a six-year-old asked you how many species of zebras exist. Most people would be satisfied with an answer they found in Wikipedia because they know there's a more than decent chance it'll be accurate and even if it's wrong, a wrong answer will do no particular harm. However, if you were asked the same question by a zoology professor, then you might want to seek the answer from multiple scholarly sources, as taxonomists may disagree on the number of species or have disputes about whether there are subspecies, and your professor is going to expect a more erudite, well-researched answer than what a six-year-old will expect.

So, how concerned you should be about credibility partly depends on why you need the information, how you are using the information, how important accuracy is, and how dire the repercussions are for being inaccurate. And realistically, how much effort you'll expend judging the credibility of a source also depends on the amount of time you have.

To sum up, what should science and engineering students know about credible sources?

1. A source is considered credible if it's reasonable to assume it will provide accurate and reliable information.
2. A source can show signs of credibility, but still contain inaccurate, unreliable information.
3. Scholarly sources (such as journal articles, conference papers, and scholarly books) are considered to be credible sources.
4. You can assess the credibility of non-scholarly sources by exploring various attributes described in this chapter.
5. You should be especially wary of the credibility of sources that discuss controversial, political, or highly emotional topics.
6. You should not assume the top result you get in a Google search is the most credible.
7. As you become a seasoned expert, your focus will shift from judging the credibility of a source to judging the rigor and utility of the content delivered in a source.

What's Next?

You've just finished the last chapter of Part III! **Part III** was all about recognizing key qualities of sources—such as scholarly, popular, primary, secondary, and credible. Next, we'll begin **Part IV**, which is short and consists of just two chapters about research assignments and research topics. The next chapter will help you interpret **research assignment instructions**.

PART IV
UNDERSTANDING THE SOURCES YOU NEED FOR RESEARCH ASSIGNMENTS

CHAPTER 18
DECODING RESEARCH ASSIGNMENT INSTRUCTIONS

Do you really need a whole chapter to help you understand class assignment instructions? Sadly, yes—but it's a short chapter! When instructors give out assignments that involve finding sources, you'll notice there's no consistency in the way they refer to sources or source qualities in their assignment instructions. This isn't just a problem in science and engineering disciplines, it's a problem across all academic departments on college campuses. The lack of standardized terminology leads to confusion, frustration, and poorer grades for students. This chapter will give you the knowledge and resources to deal with this issue in *any* course you take, whether the course is about science, engineering, literature, history, or whatever.

Assignment Instructions Can Be Really Confusing

In at least some of your courses, you'll undertake research assignments that will involve finding sources. Your instructor will likely provide instructions stipulating the types of sources they want you to find and how many. Sometimes these instructions are fairly vague and open-ended—for example, your instructor might tell you to find and cite some credible sources. Sometimes the instructions are more detailed and explicit—for example, your instructor might tell you to find and cite a certain *number* of information sources and stipulate those sources have to be of a certain *type*. That is, you might be told to "Find and cite at least five peer-reviewed sources."

How vague or how detailed the assignment instructions are depends on the particular instructor's teaching style and their learning objectives for the assignment.

The problem is that different instructors have different ideas about what constitutes a credible source, a peer-reviewed source, a scholarly source, etc. So, even though you learned in previous chapters what words like "peer-reviewed," "scholarly," and "credible" mean, you can't assume your instructors are using the same definitions.

Furthermore, instructors use lots of different terminology to refer to sources—often interchangeably. For example, as you learned in Chapter 6, instructors might variously refer to journal articles as "studies," "articles," "papers," or "the literature." So, you can expect to encounter lots of different and confusing instructions stipulating which kinds of sources you're supposed to use. Every instructor seems to have their own preferred terms and definitions!

Clarify Your Research Assignment's Source Requirements

To ensure you fully grasp what sources you're required to use for your research assignment, you'll almost certainly have to ask your instructor for clarification. Asking questions and getting clarification is *always* a good idea—it'll help you better understand the preferred terminology, definitions, and research expectations across different disciplines. Asking questions also helps your instructor appreciate how challenging it is for students to decode the wide variety of instructions they hear from different instructors across different disciplines.

Use Table 18.1 to help articulate your questions. If your instructor doesn't understand why you're asking for clarification, then feel free to show them this table. Seeing it should help them understand how and why our lack of standard vocabulary creates so much confusion.

Table 18.1

Typical source instructions (Within the brackets are different wording variations you might encounter from different instructors.)	*What to clarify with your instructor*
"Use [**peer-reviewed** or **refereed**] [**sources** or **literature** or **research**]."	Ask whether you should use *only* **journal articles** - and whether that includes both **research articles** and **review articles**. You may want to ask if **scholarly books** and **conference papers** are also allowed, even though it's usually not possible to confirm whether those sources are peer-reviewed. If you like, ask about the use of **dissertations** and **theses** as well.
"Use [**scholarly** or **academic**] [**sources** or **literature** or **research**]."	Confirm that you're allowed to use **journal articles** (both **research articles** and **review articles**), as well as **conference papers** and **scholarly books**. You may also want to check if you're allowed to use **dissertations** and **theses**.

Continued

Table 18.1 *Continued*

"Use [**peer-reviewed** or **refereed** or **scholarly** or **academic**] [**journals** or **studies** or **articles** or **papers**]."	Ask whether you should only use **research articles** and **review articles** from journals, and check whether **conference papers** are also allowed.
"Use [**empirical** or **primary** or **research** or **original**] [**journals** or **studies** or **articles** or **papers**]."	Confirm this means you should only use **research articles** from journals, and check whether **conference papers** are also allowed.*
"Use **credible sources**."	Ask your instructor for examples of the types of sources they consider to be credible and/or ask them what qualities they want you to look for to distinguish credible sources from those that are not.
"Use **popular sources**."	Ask the instructor whether this includes **websites**, **popular books**, **news**, and **magazine articles**—and if there are additional types of sources they would include in this category.
"Use ________."	There are lots of other types of instructions you might hear. When in doubt, ask for clarification.

*Remember from Chapter 16 that what counts as a primary source depends on the discipline. If you're in a non-science class and asked to use primary sources, then ask your instructor for *their* definition of "primary sources."

In addition to the source instructions you see in Table 18.1, you may be told to find and use more specific types of sources, such as journal articles that follow a particular **study design**. This is fairly common in assignments given out in health sciences, psychology, or sociology classes. For instance, you might be told to find and cite randomized, controlled, clinical trials, or longitudinal studies, etc. Or you might be told to find and use **review articles**—and possibly specific types of review articles like **systematic reviews**. Refer to Chapters 7 and 8 for more information about these types of sources, and don't be afraid to ask your instructor for additional clarification if you need it.

Does Your Assignment Tell You How to *Find* Sources?

Sometimes, instead of explicitly telling you what *types* of sources you should use, instructors will tell you how to go about *finding* sources. Table 18.2 tells you how to decode those instructions.

Table 18.2

Typical "finding" instructions	*What to clarify with your instructor*
"Use [**specific database**] to find sources."	Ask if you're allowed to use *any* type of source that [**specific database**] retrieves on your topic, or whether you're only supposed to use certain types of sources, such as journal articles, conference papers, scholarly books, etc.

Continued

Table 18.2 *Continued*

"Use the library's website to find sources."* or "Use library databases to find sources."*	Ask if you're allowed to use any type of source that you find while searching library databases or whether you're only supposed to use certain types of sources, such as journal articles, conference papers, scholarly books, etc.

*When instructors provide you with instructions to use the library's website or library databases to find sources, they don't realize why these instructions are confusing. To help them understand, explain that many of the databases offered by your campus library will retrieve results from both popular and scholarly sources. So, that's why you need to clarify if any source (regardless of whether it is popular or scholarly) is fair game to use.

Or, Maybe Your Assignment Tells You How *Not* to Find Sources?

Instructions for how *not* to find sources can be especially confusing and ambiguous. See Table 18.3 for advice on how to get clarification.

Table 18.3

Typical restrictive instructions	*What to clarify with your instructor*
"Don't use websites as sources." or "Don't use dot.com websites as sources."	When instructors say these things, they usually mean they don't want you to use websites as a *type of source* (a.k.a. website-sources).* But, you'll need to clarify and prompt them to let you know if you can use traditional sources (like journal articles, conference papers, etc.) that you find on the internet.
"Don't use Google to find sources." or "Don't search the internet to find sources."	Again, this usually means the instructor doesn't want you to use websites as a *type of source* (a.k.a. website-sources).* But, you'll need to clarify. Ask your instructor *how* they want you to go about finding sources and ask them to spell out exactly which types of sources you're allowed to use.

*Refer to Chapter 11 for more information about website-sources.

You can help your instructors understand why these instructions are so confusing. Let them know that your campus library delivers most of their resources through their homepage, which is itself a website! Plus, most of the full-text scholarly content linked from your library's website is delivered from **.com** websites. (A good example is the journal *Nature*, which is available from the website **nature.com**.) So, to forbid you to use content from websites and/or **.com** websites is severely limiting; nor is it clear what purpose this would serve.

Understanding Your Assignment's Source Requirements Will Lead to Better Grades and Less Frustration

When you work on a research assignment, typically part of your grade will be based on whether you properly followed the assignment's source requirements. That's why you need to fully understand those instructions! But, there's another reason why it's critical to understand the sources you're supposed to use. Suppose your assignment instructions indicate that *you* are responsible for choosing a topic to research and write about. If so, you need to choose your topic *based on the sources you're supposed to use.* If you don't, you'll find the research process really frustrating. Don't worry, the next chapter will thoroughly explain this issue.

To sum up, what do science and engineering students need to know about decoding assignment instructions?

1. Instructors are inconsistent in the terms they use to refer to sources; nor are they consistent in how they define source qualities like scholarly, peer-reviewed, credible, etc.
2. Use the tables in this chapter to get clarification from your instructor on how *they* define source qualities, so you fully understand the exact types of sources you're supposed to use.

What's Next?

In the next chapter we'll look at strategies for **picking a good topic** that's appropriate for the types of sources you're supposed to find and use.

CHAPTER 19
CHOOSING A TOPIC THAT MATCHES SOURCE REQUIREMENTS

Choosing a good research topic will save you a lot of time and agony. One of the most common mistakes students make is choosing a topic that doesn't match up well with their assignment's source requirements, but it's also common for students to choose topics that are too broad or too narrow. This chapter will help you avoid those problems. You'll also learn the importance of testing your topic to make sure you can find a reasonable number of relevant, appropriate sources for it.

What Does It Mean to "Choose a Topic"?

When you're given an assignment to find sources on a topic, your instructor may provide you with a specific topic to research, or give you a choice of topics to pick from, or allow you to choose your own research topic. If you're told to choose your *own* topic to research, your instructor may want you to state it in the form of a research question or a hypothesis. Selecting your own topic may sound pretty easy, but it can be rather tricky. Students frequently pick topics that are either too broad or too narrow and this can make the research process frustrating. It's also common for students to pick topics that don't align well with the types of sources the instructor wants you to use. So, if you want your research process to be smooth, interesting, and gratifying, then you need to be strategic about selecting a topic.

Choose a Topic that Aligns with Your Research Assignment's Source Requirements

Use the tips from the previous chapter to ensure you fully understand the sources your instructor wants you to use for your research assignment. Then—and only then—should you start thinking about a topic, and you'll want to choose one that aligns well with your source requirements. Why? Let's look at an example.

Suppose you're given a research assignment where you have to find and cite at least five peer-reviewed journal articles. As you know from Chapter 14, journal articles are a type of scholarly source, and scholarly sources discuss scholarly topics. Recall that scholarly topics include biology, medicine, sociology, engineering, and any other discipline represented by academic departments on a college campus. Let's say you chose to research the culture of motorcycle gangs and the role of women in those gangs. This is a sociological topic, and sociology is a scholarly discipline, so you're likely to find scholarly sources, like journal articles, on this topic.

What if instead you chose to research motorcycle maintenance and repair? That's not a scholarly topic; therefore, it won't be discussed in scholarly sources. You'll have a very hard time finding journal articles on that topic. That's why you need to ensure your topic aligns well with the types of sources you need to find and use. When you do, the research process is much easier and more rewarding. So, instead of researching motorcycle maintenance and repair, pick a topic that scholars, like engineers, would be likely to study, such as "What factors improve the performance and emissions of motorcycle engines?" You're much more likely to find scholarly sources examining that issue.

Make Sure Your Topic Is Not Too Broad

Topics that are too broad can be difficult to research and write about because there's simply too much to cover. If your topic is a single word or a simple phrase, it's too broad, like these topics:

- **bioengineering**
- **food safety**
- **climate change**
- **prairie dogs**

Another problem with broad topics is that when you write about them, your paper will end up sounding like a really long definition or overview, like an encyclopedia entry. For college- and graduate-level assignments, instructors usually want you to investigate an issue or problem rather than just defining something. A topic that undertakes an investigation or analysis—or that explores a question or hypothesis—will also be more interesting to write about.

But sometimes it's hard to get from a broad topic to a narrower, more interesting topic that investigates a particular issue. If you're having trouble, think about your topic in terms of a specific problem that needs to be explored and try adding in parameters that could give the topic more focus. Here are some questions to ask yourself that might help you narrow your topic:

- **What** problem interests you about this topic?
- **Who** is involved or affected?
- **When** is this topic relevant—right now, or is it more historical?
- **Where** is the topic relevant? Consider location (such as the United States, Africa, the Pacific Northwest, rural areas, etc.) and/or contexts (such as elementary schools, prisons, corporations, etc.)

Let's look at a couple of examples. If you think about the broad topic "prairie dogs" in terms of an issue or problem, and then add in a **where** parameter, you might land on this narrower topic:

- **What factors contribute to the success or mortality of urban prairie dog colonies?**

And, if you look at the issue of "food safety" in terms of an investigation that specifies **who** is affected, then you might come up with this narrower topic:

- **How effective are federal regulations for food safety in protecting Americans' health and well-being?**

See how these narrower topics are framed so that you're looking for answers to a specific problem? See how much more interesting it would be to research these topics?

You can also turn topics into a hypothesis by stating them as a sentence for which an investigation will reveal an answer, like this:

- **Urban prairie dog colonies are at greater risk of epidemic disease then rural colonies.**

You might be thinking you'd rather stick with a broad topic because it'll be easier to find an abundance of sources, and that'll make the research process easier. But having lots of sources to choose from can actually make the research and writing process more difficult. Let's look at an example. Suppose you're an engineering student who has chosen to research the topic **water purification.** Your assignment stipulates that you need to find and cite five peer-reviewed journal articles. Figure 19.1 shows you some records for articles that might come up in a database search on this topic.

1. Solar Water Purification: High Yield and Low-Cost Solar Water Purification via Hydrogel-Based Membrane Distillation
Get full text
Authors: Lu, Hengyi, et al.
Source: *Advanced Functional Materials*, 2021, Vol. 31 (19) 2170135

2. Water purification: oil-water separation by nanotechnology and environmental concerns
Get full text
Authors: Lee, Chee Huei, et al.
Source: *Environmental Science. Nano*, 2017, Vol. 4 (3), pages: 514-525

3. Water purification using magnetic assistance: A review
Get full text
Authors: Ambashta, Ritu D., et al.
Source: *Journal of Hazardous Materials*, 2010, Vol. 180 (1), pages: 38-49

4. Fabrication of a novel thin-film nanocomposite (TFN) membrane containing MCM-41 silica nanoparticles (NPs) for water purification
Get full text
Authors: Yin, Jun, et al.
Source: *Journal of Membrane Science*, 2012, Vol. 423-424, pages: 238-246

5. Graphene oxide-silver nanoparticle membrane for biofouling control and water purification
Get full text
Authors: Sun, Xue-Fei, et al.
Source: *Chemical Engineering Journal*, 2015, Vol. 281, pages: 53-59

6. Gravity-Driven Hybrid Membrane for Oleophobic-Superhydrophilic Oil-Water Separation and Water Purification by Graphene
Get full text
Authors: Yoon, Hyun, et al.
Source: *Langmuir*, 2014, Vol. 30 (39), pages: 11761-11769

7. Apple Peels: A Versatile Biomass for Water Purification?
Get full text
Authors: Mallampati, Ramakrishna, et al.
Source: *ACS Applied Materials and Interfaces*, 2013, Vol. 5 (10), pages: 4443-4449

8. Water Purification and Microplastics Removal Using Magnetic Polyoxometalate-Supported Ionic Liquid Phases (magPOM-SILPs)
Get full text
Authors: Misra, Archismita, et al.
Source: *Angewandte Chemie*, 2020, vol. 59 (4), pages: 1601-1605

Figure 19.1

Each of these articles is about water purification, and there are *hundreds* more besides these. Because your topic is so broad, any of these articles would be relevant to use and cite. Again, it might sound great to have so many relevant sources to choose from, but really—it's not. Notice how the sources shown in Figure 19.1 cover lots of different techniques for tackling water purification problems using solar power, nanotechnology, magnets, gravity, biomass, etc. Furthermore, some are about perfecting membranes, others are about what impurities to purify (such as microplastics), and some are about the environmental impacts of water purification—and this is just within the first eight results of your search! How could you pull together a coherent paper that talks about water purification from all these different angles? It would be pretty frustrating. Furthermore, choosing which results to use from all these equally relevant options is going to be a random and meaningless process.

Another problem with broad topics is that there can be a mismatch between the broadness of your topic and the narrowness of the sources you're supposed to use. Look at how narrowly focused most of the journal articles are in the results shown in Figure 19.1. As you know from Chapter 7 this is typical; journal articles (and research articles in particular) focus on a very narrow, well-defined research project. It's going to be frustrating to write a paper that provides an overview of the topic "water purification" if your assignment requires that you find and use very narrowly focused studies, such as journal articles.

To narrow this topic, use the techniques you just learned to focus on a more specific *problem* in a more specific *context*. For example, you might want to examine a particular type of water that needs purifying, such as wastewater, and even zero in on certain types of contaminant removal, such as pharmaceutical residues. You might even be interested in leveraging sustainable technologies to do this, such as solar technology. Your new, more focused topic could be:

- **How is solar technology being leveraged to remove pharmaceutical contaminants from wastewater?**

You'll find fewer but more targeted results on this more specific topic, as you see in Figure 19.2. Also notice that your narrower topic aligns better with the types of sources you are supposed to find and use (journal articles).

1. Decontamination industrial pharmaceutical wastewater by combining solar photo-Fenton and biological treatment
Get full text
Authors: Sirtori, C., et al.
Source: *Water Research*, 2009, Vol. 43 (3), pages: 661-668

2. Solar photo-Fenton oxidation for the removal of ampicillin, total cultivable and resistant E. coli and ecotoxicity from secondary-treated wastewater effluents
Get full text
Authors: Ioannou-Ttofa, Lida, et al.
Source: *Chemical Engineering Journal*, 2019, Vol. 355, pages: 91-102

3. Environmental sustainability of the solar photo-Fenton process for wastewater treatment and pharmaceuticals mineralization at semi-industrial scale
Get full text
Authors: Foteinis, Spyros, et al.
Source: *The Science of the Total Environment*, 2018, Vol. 12, pages: 605-612

4. Elimination of pharmaceutical pollutants by solar photoelectron-Fenton process in a pilot plant
Get full text
Authors: Bugueno-Carrasco, Sebastian, et al.
Source: *Environmental Science and Pollution Research International*, 2021, Vol. 28 (19), pages: 23753-23766

5. Optimization of pharmaceutical wastewater treatment by solar/ferrioxalate photo-catalysis
Get full text
Authors: Monteagudo, J. M., et al.
Source: *Journal of Environmental Management*, 2013, Vol. 128, pages: 210-219

Figure 19.2

Another Way to Narrow Down Broad Topics

There's another useful technique for narrowing down broad topics, and that's to start searching some databases to get ideas for how to narrow. To use this technique, you'll need to know a little bit about databases, so quickly skim Chapter 20 to learn about **discovery tools** and **article databases** and how to find them on your campus library's website. Or, you can simply search Google Scholar if you prefer.

Enter a few keywords into the database that represent a broad topic that interests you. Suppose you're interested in researching **biosand water filters**, so you run a search for those keywords in Google Scholar. Then, scan over your results to get ideas for how you might narrow your topic. For example, the first result shown in Figure 19.3 might give you ideas about focusing your topic on the efficacy of biosand filters versus ceramic filters, whereas the second result might make you think about narrowing your topic to explore how grain size affects the performance of biosand filters.

The effect of increasing grain size in **biosand** water **filters** in combination with ultraviolet disinfection
TE Frank, ML Scheie, V Cachro... - **Journal of water** ..., 2014 - iwaponline.com
... the source as **Journal of Water**, Sanitation and Hygiene for Development 4(2), 206–213. ...
purification processes in this study (**biosand filtration** and UV disinfection) for each BSF. ...
☆ Save 99 Cite Cited by 13 Related articles All 6 versions

A critical evaluation of the appropriateness of **ceramic** and **biosand filters** in rural Cambodia
HM Murphy - 2010 - atrium.lib.uoguelph.ca
... Currently, there are two water treatment technologies that have been widely implemented across Cambodia: **ceramic** water **filters** and **BioSand filters**. Both technologies have proven to ...
☆ Save 99 Cite Cited by 8 Related articles All 3 versions »

Figure 19.3

So, then you might come up with a topic like: *How can biosand water filter performance be optimized by grain size?* After that, you'll want to test your topic to make sure you can find enough relevant sources. You'll learn more about testing topics shortly, but to get an idea what to do, just enter keywords representing your new, narrower topic into one or more databases to see if you're getting enough relevant results to write a paper.

Make Sure Your Topic Is Not Too Narrow

It's possible to choose a topic that's so narrow, obscure, or weird that there are no sources about it—or, too few to write a research paper. Here are some examples:

- **Does dark sky community status boost tourism in Flagstaff, Arizona?**

- **Which types of roller-coaster seats and restraints cause the fewest injuries?**
- **What efforts have been made to tame hummingbirds?**

Topics can also be too narrow in a different way. Consider the topics below that ask questions for which the answer will be so short and succinct that it would be impossible to write enough to fill up a whole research paper:

- **What percentage of the Minnesota population is Native American?**
- **What is the latest standard for measuring pulse velocity through concrete?**
- **Which diseases are the leading causes of death in the United States?**

You'll also want to avoid landing on a topic for which the information you need is likely to be proprietary, meaning it's not shared outside the company that created it. Below are a few examples:

- **What are the best methods and materials for creating artificial heart valves?**
- **How was ArcGIS software developed and marketed?**
- **How does Bristol-Myers Squibb organize its research teams to develop new pharmaceuticals?**

To see how to broaden a topic, consider this example topic provided at the beginning of this section:

- **Does dark sky community status boost tourism in Flagstaff, Arizona?**

This is a very narrow topic that's restricted to a very specific location and it's unlikely you'll find any articles on it. Not only that, if you're interested in how dark skies promote tourism, why restrict yourself to investigating just one town? A better topic would be:

- **Do efforts to preserve dark skies promote tourism?**

You are much more likely to find articles on that topic, and any articles you find will likely have information applicable to Flagstaff, or any other community that protects its dark skies.

Try to State Your Topic as a Formal Question or Hypothesis

If you can state your topic as a question or hypothesis, you'll find it easier to stay focused on exactly what you need to find when you start searching for sources. Stating your topic as a sentence will also help you identify the fewest, most essential keywords that represent your topic, which is really important—and you'll learn all about how to do that in Chapter 24.

It's also a good idea to try to write your topic sentence using the most formal, professional words possible. As an example, suppose you're interested in this topic:

- **To what extent do teenagers engage in risky behavior as a result of peer pressure?**

As you start searching this topic in databases, you might notice that many authors are using more formal words and phrases like **adolescents** and **"risk taking"** instead of the keywords in your topic sentence. In that case, reword your topic sentence with the more formal words keywords you found:

- **To what extent do adolescents engage in risk taking as a result of peer pressure?**

In general, be on the lookout for the preferred terminology that scholars use like "adolescent" instead of "teenager," especially when you're searching for scholarly sources on a topic. You'll usually get better results from databases when you search using these more formal keywords.

Test Your Topic

Really, the only way to ensure you've landed on a good topic for which you can find the right kinds of sources—and not too many or too few—is to **test your topic**. Testing your topic means doing some preliminary research on it to see what you can find (see Figure 19.4).

The process of testing your topic is similar to conducting actual research on your topic—that is, you'll need to identify keywords that represent your topic, choose and search databases for sources, and determine whether your search results show a promising number of relevant sources. However, when testing a topic, you can be pretty slapdash in your searching because you're just doing a trial run. All you need to find out at this preliminary stage is whether you can find a reasonable number of appropriate sources. If so, then you can settle on that topic and begin your research process in earnest. The next several chapters will help you gain the skills you need to select databases and search them in order to test your topic to see if it's viable.

It's a good idea to start testing topics soon after you get your research assignment so you can establish a viable topic as soon as possible. Realizing that you've chosen an inappropriate, difficult-to-research topic the day before your research paper is due is frustrating and panic-inducing. Make sure you take special care to ensure your topic

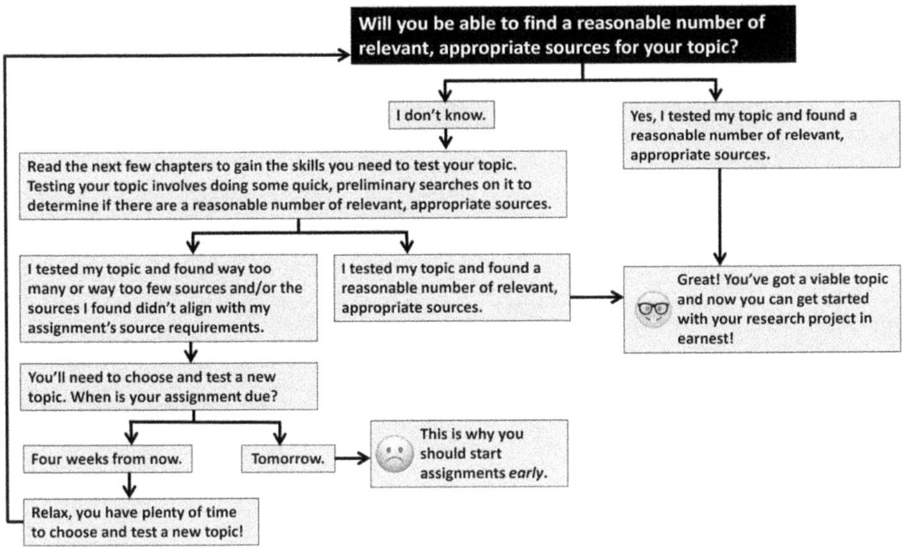

Figure 19.4

aligns well with your source requirements, otherwise you won't find many (or any!) relevant sources. If you leave it too late and don't have time to change your topic to better match your source requirements, you'll be stuck writing a paper where you have to cite inappropriate, irrelevant sources—and then contrive a way to make those sources sound relevant within your text. This makes the research and writing process feel kind of phony, and you learn a lot less from it. On the other hand, if you take the time early on to establish a good topic, the research process will be easier, as well as more fun and interesting, and you're more likely to write a good paper.

To sum up, what do science and engineering students need to know about choosing research topics?

1. Make sure you choose a topic that aligns with your assignment's source requirements.
2. Avoid topics that are too broad or too narrow, as they make research frustrating.
3. Test your topic before you settle on it!

What's Next?

You're done with **Part IV** and moving on to Part V! **Part V** is all about **databases**, including different types of databases, how they work, and how to choose the best to search. In the next chapter, you'll learn about different types of databases.

PART V
ALL ABOUT DATABASES

CHAPTER 20
TYPES OF DATABASES

Bewildered by the array of databases available? This chapter will break it down for you. You'll learn about the most important databases used by scientists and engineers, and the types of sources those databases provide. You'll also learn how databases differ in scope and subject matter, as well as how to find databases on campus library websites.

Types of Databases

A **database** is a collection of related information, in an electronic format, that serves the dual purpose of storing that information and making it findable. There are many different types of databases that serve different functions. For example, Internet Movie Database (IMDb) provides information about movies and TV shows. Your online credit card account is a database of your credit card transactions. Google Scholar is a database for finding published sources, such as scholarly journal articles, conference papers, books, etc. Your campus library provides even more databases for finding published sources. It's these databases for finding published sources that we'll focus on in this chapter.

Before we get started, be aware that the purpose of most databases discussed in this chapter is to help you find *records* that represent published sources. Within the record there's usually a link, or some other information, that will lead you to the *actual source.* For example, a database record might contain a link leading you to the electronic full text of the source, or it might contain information about where to find a print book on the shelves in your library. You'll learn more about database records and the information they contain in the next chapter, and you'll learn how to get to the full text of a source in Chapter 28.

Now let's look at the different types of databases that campus libraries typically offer for finding sources. We can categorize them into six main types:

1. A **catalog** for finding books and other media owned by a particular library
2. **Article databases** for finding articles (such as journal articles, conference papers, newspaper articles, and magazine articles)
3. **Content-specific databases** for finding one specific type of content—such as data, patents, videos, newspaper articles, etc.
4. **Publisher databases** for finding sources produced by a particular publisher

5. **Pre-print databases** for finding manuscripts of research articles and papers
6. A **discovery tool** that can retrieve records from a variety of other databases

Sounds like a lot of different databases, right? Well it is, but don't worry because most scientists and engineers only use a few types of these databases. For example, a biomedical scientist might use the **article database** PubMed for 95 percent of their needs. And, if you plan to become an electrical engineer, you may find the **publisher database** IEEE Xplore serves most of your needs. As you ease into your chosen profession, you'll learn which databases are most important for that profession. So, as you read about the different databases in this chapter, don't try to keep track of all of them. Instead, just use this chapter to get a general sense of the variety of databases out there and how they differ. Let's begin with catalogs.

1 Catalogs

To find out what books your campus library has to offer, you'd search a database called a "catalog." A **catalog** stores data on the unique set of books owned by a particular library. Catalogs also contain records for other materials owned by a library, such as video recordings, dissertations, theses, maps, etc. Some campus libraries offer a way to search their catalog right on their homepage, like you saw in Figure 3.1, where you'd click a tab to search the catalog. In the example in Figure 20.1, from a different library, there's a link right on the home page to "Search Catalog."

Figure 20.1

Many libraries don't offer *any* option labeled "catalog" on their homepage. For example, the campus library homepage shown in Figure 20.2 has no links or tabs to a "catalog," but instead offers a drop-down menu to limit your search to books and media. This is essentially a search of that library's catalog, even though the word "catalog" is not used.

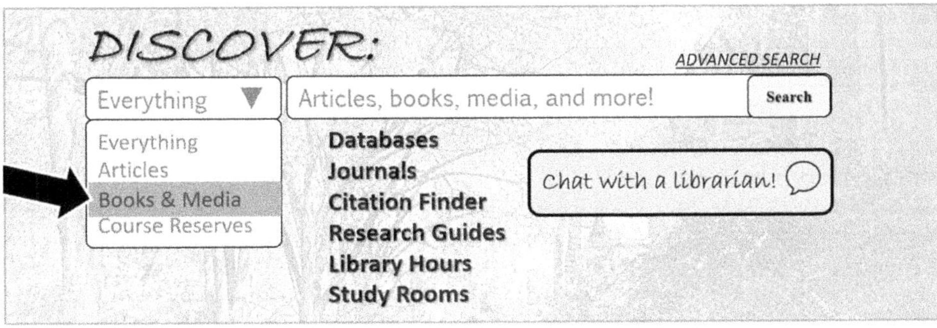

Figure 20.2

Some libraries don't provide any of the options you just saw, but instead provide a **discovery tool** for library users to search. This type of database can be configured to retrieve records for items owned by a particular library, just like a catalog would do. If you want to learn how to use a discovery tool as a catalog, it's spelled out in detail in Chapter 3—just look for the section on "Finding books." As for discovery tools, you'll learn more about how they work a bit later in this chapter.

To sum up, libraries are all over the map when it comes to how they present their catalog, and many libraries no longer use the word "catalog" at all. Not only that, the way to navigate to the catalog and search it is also variable. Sometimes it's a link, sometimes it's found in a drop-down menu, sometimes it's on a tab, sometimes it's embedded in a discovery tool search. Why don't all libraries provide an easy-to-find direct link to search their catalog? Because over the years there's been a growing concern that students don't know what the word "catalog" means. So, many libraries have eliminated this database name from their website and opted for other ways to present users with access to their local collections. Not surprisingly, different libraries have landed on different solutions, which means you won't encounter consistent terminology or navigation from one library website to the next. If it's not clear how to search the catalog on *your* library's homepage, then ask library staff for help.

One more thing to mention about catalogs is that they weren't always databases. Before digital technology, the items in a library's collection were only searchable by rifling through cabinets with little drawers containing cards that represented each item the library owned (called a **card catalog**). Each card provided a location and call number for finding a particular item on the library's shelves. A **call number** is like an address, it tells you exactly where to find the book on the shelf. After libraries digitized their card catalog records, they referred to the resulting database as a "catalog," rather than a "card catalog," because physical cards were no longer necessary. For a summary of what a catalog is, see Table 20.1.

Table 20.1

Database	*Subject Coverage*	*Source Types Tracked*
Catalog	Multidisciplinary	Books owned by a particular library—plus videos, maps, and other media owned by that library

WHERE TO FIND THE CATALOG: Look on your library's homepage for links, tabs, or drop-down menu options to search the "catalog"—or it might be labeled "books" or "find books" or something similar. Alternatively, search your library's discovery tool, then refine your search to *books only*. If you can't figure out how to find and search your library's catalog or discovery tool, then ask library staff for help.

2 Article Databases

Not surprisingly, **article databases** help you find articles. Examples of databases for finding articles include PubMed, Web of Science, Scopus, Academic Search, etc. Article databases like these are often referred to as just **databases**—but sometimes they're called **bibliographic databases, library databases**, or **index databases**. The name "bibliographic database" comes from the fact that these databases are populated with bibliographic data. **Bibliographic data** include all the little bits of data that represent a published source, such as the author, title, publication date, etc. The name "library database" comes from that fact that you typically access these databases from a library's website. But where does the name "index database" come from? Many years ago, article databases only existed in print form as a set of volumes called an **index**. An article database is really just a digitized version of a print index, so that's the origin of the name "index database."

A link to navigate to article databases is usually right on the homepage of your campus library's website, like you saw in Figure 6.7. But each library is different. For example, in the library website depicted in Figure 20.1 the link is just to the right of "Search Catalog" and is called "Find Databases." In Figure 20.2 the link is simply called "Databases." As you can see, the name of this link varies from library to library and could be any of the following, or something similar:

- **Databases**
- **All Databases**
- **Databases A–Z**
- **Find Databases**
- **Research Databases**
- **See All Databases**

When you click the link to view your library's databases, you'll usually get a page that allows you to view them in a variety of ways. In the example shown in Figure 20.3, you can:

1. Search for a particular database by name or keyword,
2. View databases by subject area,

3. View databases by content type,
4. Navigate to a database by title using an alphabetical list.

Figure 20.3

Table 20.2 provides some examples of article databases and shows the subjects they cover and the types of sources they track. Notice that some cover many different subjects and so are considered **multidisciplinary**. Others focus on particular subjects and are considered **subject-specific**. (We'll talk more about disciplinary coverage in a moment.) All these databases help you find journal articles, but some also track newspaper articles and magazine articles, and some track a few other source types to a greater or lesser extent—such as conference papers, books, etc.

Table 20.2

Database	*Subject Coverage*	*Source Types Tracked*
Academic Search (Premier or Complete)	Multidisciplinary	Journal articles, magazine articles, newspaper articles
ACM Guide to Computing Literature	Subject-specific: Computing and Information Technology	Journal articles, conference papers, books, dissertations/ theses, etc.
ADS (Astrophysics Data System)	Subject-specific: Physics, Astronomy	Journal articles, conference papers
BIOSIS	Subject-specific: Biology	Journal articles, conference papers
CAB Abstracts	Subject-specific: Agriculture, Forestry, Environmental Science, Human Health and other applied life sciences	Journal articles, conference papers
CINAHL	Subject-specific: Nursing, Allied Health	Journal articles
Compendex	Subject-specific: Engineering	Journal articles, conference papers
GeoRef	Subject-specific: Geology	Journal articles, conference papers, books, maps, etc.

Continued

Table 20.2 *Continued*

Database	*Subject Coverage*	*Source Types Tracked*
Google Scholar	Multidisciplinary	Journal articles, books
GreenFILE	Subject-specific: Environmental Science	Journal articles, magazine articles, books
ERIC	Subject-specific: Education	Journal articles, conference papers, books, curricular materials, etc.
MathSciNet	Subject-specific: Math, Computer Science, Statistics	Journal articles, books
PsycINFO	Subject-specific: Psychology	Journal articles, books
PubMed	Subject-specific: Medicine	Journal articles, books
SciFinder	Subject-specific: Chemistry	Journal articles, patents, books, conference papers, etc.
Scopus	Multidisciplinary	Journal articles, conference papers, books
SocINDEX	Subject-specific: Sociology	Journal articles
Web of Science	Multidisciplinary	Journal articles, conference papers
Zoological Record	Subject-specific: Zoology	Journal articles, conference papers, books

WHERE TO FIND ARTICLE DATABASES: Look for a link on your library's homepage that indicates it will lead you to databases. If you don't see any link like that, then ask for help.

In case you're wondering where all these article databases come from, many are produced by private companies like EBSCO, Gale, Clarivate, ProQuest, Elsevier, etc. For example, the company Clarivate produces Web of Science, whereas the company Elsevier produces Scopus. Others are produced by professional societies or organizations—such as SciFinder, which is produced by the American Chemical Society and MathSciNet, which is produced by the American Mathematical Society. A few databases are produced by United States government agencies, such as PubMed, which is produced by the National Library of Medicine. And of course, Google produces Google Scholar. Those last two examples (PubMed and Google Scholar) are freely available for anyone to search, but most article databases require a subscription to access. Campus libraries pay these subscription fees so that you don't have to!

3 Content-Specific Databases

Content-specific databases help you find just one specific type of content. For example, there are content-specific databases devoted to helping you find just patents, or just newspaper articles, or just videos, etc. Suppose you need to search for patents and you want to see what databases your library provides for doing that. Go to your campus library's homepage and click the same link to "databases" that you'd use for finding article databases. After that, look for a heading or menu that allows you to view databases by

content, **format**, **type**, or **source type** (the exact wording varies on different library websites). Then, see if there's an option to view patent databases, like you see in Figure 20.4.

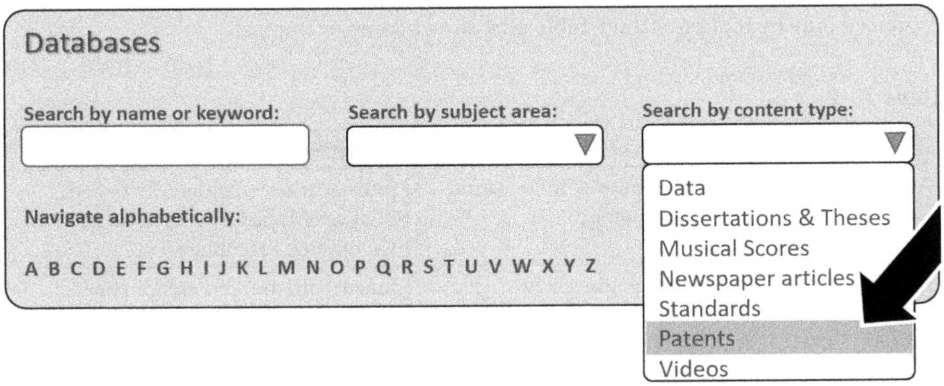

Figure 20.4

As all library websites are different, your library may not provide a way to navigate to content-specific databases (like you see in Figure 20.4) or your library may present lists of content-specific databases elsewhere on their website. If you are looking for content-specific databases and you can't figure out how to navigate to them, then ask library staff for help. Some examples of content-specific databases are shown in Table 20.3.

Table 20.3

Database	*Subject Coverage*	*Content Type Tracked*
Access World News	News	Just newspaper articles
Academic Video Online (AVON)	Multidisciplinary	Just videos
ASTM Compass	Engineering	Just standards
Data Citation Index	Multidisciplinary	Just data
Derwent Innovations Index	Engineering, chemistry, etc.	Just patents
Gale OneFile: Popular Magazines	News/multidisciplinary	Just magazine articles
Gale OneFile: News	News	Just newspaper articles
Google Patents	Engineering, chemistry, etc.	Just patents
ICPSR	Social sciences	Just data
ProQuest Dissertations and Theses Global	Multidisciplinary	Just theses and dissertations
ProQuest Newspapers	News	Just news articles
TechStreet	Standards	Just standards

WHERE TO FIND CONTENT-SPECIFIC DATABASES: You can usually find these databases by using the same link for finding article databases that's provided on your campus library's homepage. Once you click that link, look for ways to view databases by content, format, type, or source type – or search for them by title.

4 Publisher Databases

Publisher databases are provided by major publishers, and they present sources produced *only* by that publisher. Table 20.4 shows some examples.

Table 20.4

Database	*Subject Coverage*	*Source Types Tracked*
ACM Digital Library	Computing, Information Technology	Journal articles, conference papers, books published by the Association for Computing Machinery
ASCE Library	Civil engineering	Journal articles, conference papers, books, etc., published by the American Society of Civil Engineers
IEEE Xplore	Electrical engineering	Journal articles, conference papers, books, etc., published by the Institute of Electrical and Electronics Engineers
Sage Premier	Any topics that Sage journals happen to cover	Journal articles published by Sage
ScienceDirect	Any topics Elsevier journals and books happen to cover	Journal articles and books published by Elsevier
SPIE Digital Library	Optics, Photonics	Journal articles, conference papers, books published by the Society of Photo-optical Instrumentation Engineers
Springer Link	Any topics Springer journals and books happen to cover	Journal articles and books published by Springer
Wiley Online Library	Any topics Wiley journals and books happen to cover	Journal articles and books published by Wiley

WHERE TO FIND PUBLISHER DATABASES: You can usually find these databases by using the same link for finding article databases that's provided on your campus library's homepage. Once you click that link, search for a specific publisher database by title.

There are also a few publisher databases that provide access to the collected output of groups of smaller publishers. Table 20.5 shows some examples.

Table 20.5

Database	*Subject Coverage*	*Source Types Tracked*
BioOne	Biology, Ecology, Environmental Science	Journal articles published by a variety of small-scale publishing companies
JSTOR	Multidisciplinary	Journal articles (and some books) published by a variety of small-scale publishing companies

WHERE TO FIND MULTI-PUBLISHER COLLECTIONS: You can usually find these databases by using the same link for finding article databases that's provided on your campus library's homepage. Once you click that link, search for a specific database by title.

5 Pre-Print Databases

Pre-print databases provide records for pre-prints, but that won't mean much to you unless you know what a pre-print is. A **pre-print** is an electronic *manuscript version* of a journal article or conference paper that's not yet been peer-reviewed. Basically, it's the manuscript that an author sends off to a publisher to get published. Some pre-print databases also contain post-prints. A **post-print** is a manuscript *after* it's been peer-reviewed and accepted for publication. Some people refer to pre-prints and post-prints collectively as **e-prints.** Likewise, pre-print databases are sometimes called e-print databases. These databases might also be called:

- **pre-print or e-print repository**
- **pre-print or e-print server**
- **pre-print or e-print archive**

Because pre-prints are manuscripts, they look different from formally published articles insofar as they've not been formatted to look like the typical PDF version of a journal article or conference paper. So, they don't look like what you see in Figure 7.1. Instead, they look more like a research paper you'd turn in to a professor with text in 12-point font and double-spaced.

What is the purpose of pre-print databases? The Wikipedia article on **Preprints** (https://en.wikipedia.org/wiki/Preprint) offers an excellent explanation:

> Publication of manuscripts in a peer-reviewed journal often takes weeks, months or even years from the time of initial submission, owing to the time required by editors and reviewers to evaluate and critique manuscripts, and the time required by authors to address critiques. The need to quickly circulate current results within a scholarly community has led researchers to distribute documents known as preprints, which are manuscripts that have yet to undergo peer review. The

immediate distribution of preprints allows authors to receive early feedback from their peers, which may be helpful in revising and preparing articles for submission.[14]

The most well-known pre-print database is **arXiv**. arxiv began as a pre-print database serving researchers in physics. Scholars in physics have a history of valuing the open sharing of information and so pre-prints have a strong appeal for them. But over time arXiv has grown and expanded into more subject areas. According to the arXiv website, this database now accepts papers in "the fields of physics, mathematics, computer science, quantitative biology, quantitative finance, statistics, electrical engineering and systems science, and economics." (From: https://arxiv.org/about/.)

Pre-print databases are usually developed, maintained, and hosted by universities or organizations. For example, here's what arXiv says about its operations:

> Operations are maintained by the arXiv Leadership Team and arXiv staff at Cornell. Governance of arXiv is led by the Leadership Team with guidance from the arXiv Scientific Advisory Board and the arXiv Member Advisory Board. arXiv is a community-supported resource funded by Cornell University, the Simons Foundation, member institutions, and donors.

Some colleges and universities operate their own small-scale pre-print databases that store the scholarly output of faculty and students at that particular college or university. These pre-print databases are usually called **institutional repositories** (or **IRs**) and they may contain other content in addition to pre-prints, such as media, curricular materials, posters presented at conferences, etc.

On some library websites you'll find links to well-known pre-print databases such as arXiv, while other libraries don't provide any links to these databases, but rather expect you to find and use them on your own, as they're freely available on the internet. Table 20.6 shows three examples of pre-print databases.

Table 20.6

Database	*Subject Coverage*	*Source Types Tracked*
arXiv	Physics, Astronomy, and Math, Computer science, Statistics, and more	Pre-prints/post-prints of research articles/papers
BioRxiv	Biology	Pre-prints of research article/papers
SocArXiv	Social science	Pre-prints/post-prints of research article/papers

WHERE TO FIND PRE-PRINT DATABASES: If your library provides links to pre-print databases, you can usually find them using the same link for finding article databases that's provided on your campus library's homepage. Once you click that link, search for a specific database by title. You can also search for lists of pre-print databases using Google, or consult these lists:

https://en.wikipedia.org/wiki/List_of_pre-print_repositories/
https://doapr.coar-repositories.org/repositories/

6 Discovery Tools

If your library offers a discovery tool, it's usually represented by a search box right on your library's homepage, and it typically invites you to search for articles, books, media, and more (or similar wording). You saw an example in Figure 3.1. There are two more examples in Figures 20.1 and 20.2; on these library home pages the discovery tools have been given names—one is called *OneSearch* and the other is called *Discover*.

If your campus library presents a search box on the homepage but calls it a catalog, then that's a different type of database, it's *not* a discovery tool. For example, take a look at Harvard Library's homepage (Figure 20.5).

Figure 20.5

If you entered a search in this search box and then clicked "Search Hollis," you're *not* searching a discovery tool—rather, you're searching the Harvard Library catalog. How do you know that? Just below the search box you're told that "Hollis is the library catalog." If you're looking at your library's homepage and unsure whether it offers a discovery tool or a catalog, then ask library staff. If they are not sure what you mean by "discovery tool," then ask if the library provides a database that can search across other databases.

Now let's move on to talk about what a discovery tool is. A discovery tool is a database that doesn't contain *its own content*; rather, it finds content for you by going out to other databases, looking in those databases for records on the topic you entered, then presenting you with whatever results it found. A discovery tool can mine all of the following types of databases at once to retrieve records for you:

1. A **catalog** for finding books and other media owned by a particular library
2. **Article databases** for finding articles (such as journal articles, conference papers, newspaper articles, and magazine articles)
3. **Content-specific databases** for finding just one specific type of content, such as data, patents, videos, newspaper articles, etc.
4. **Publisher databases** for finding sources produced by a particular publisher
5. **Pre-print databases** for finding *manuscripts* of research articles and papers

However, the records that discovery tools can retrieve from these other databases are a bit random due to various technical limitations and set-up configurations. For example, depending on the library, a discovery tool may not be set up to search *any* pre-print databases. Or the discovery tool may not be able to search within certain article databases. So, these tools don't always do a thorough job of searching across databases.

Furthermore, most discovery tools are configured so that when you run a search on a topic, it won't show you every result it can find on that topic. Instead, it will just display records for sources that *your library subscribes to*. That sounds great, but actually it's a limitation. Suppose you were in a geology class working on a research project about how to predict soil liquefaction caused by earthquakes. For this topic, the following article would be really useful to consult:

> Zhang, Y., et al. (2021). The adoption of deep neural network (DNN) to the prediction of soil liquefaction based on shear wave velocity. *Bulletin of Engineering Geology and the Environment*, *80*(6): 5053–5060.

However, if you enter a keyword search such as **earthquakes soil liquefaction** into your library's discovery tool, the article above would *only* be retrieved if your library subscribes to the journal that article is published in (*Bulletin of Engineering Geology and the Environment*). So, this is a serious limitation if you are interested in finding the best, most relevant sources out there. (Note that a few campus libraries have configured their discovery tool to search beyond their local collections, but most discovery tools are not configured to do this.)

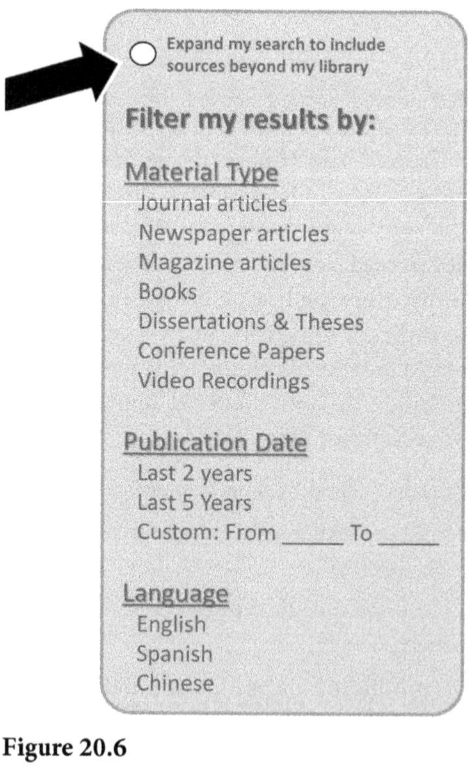

Figure 20.6

There is a way around this limitation. Most discovery tools offer an option to expand your search to retrieve additional sources your library doesn't happen to own (see Figure 20.6).

If your library's discovery tool offers this option, it usually shows up after you've run a search and it may appear above the list of ways you can refine your search. Note that the wording for this option varies from library to library.

Discovery tools are complicated and this discussion of how they work is greatly simplified, but it still gives you the gist of what's happening. These tools are useful if you are in a rush to quickly find some sources and you don't need to search thoroughly, or you don't need the tailored search features that other article databases offer (more on that in Chapter 22). The fact is that for serious researchers, discovery tools are usually not the best choice. For an overview of discovery tools, see Table 20.7.

Table 20.7

Database	*Subject Coverage*	*Sources Tracked*
Discovery tool	Multidisciplinary	Gathers all types of sources for you *from other databases*

WHERE TO FIND THE DISCOVERY TOOL: Look for a search box on your campus library's homepage that invites you to search for books, articles, media, and more—or something like that. If you don't see anything like that, then ask library staff if your library provides a database that searches across other databases. If your campus library has a search box on the homepage that's labeled "catalog," then that's *not* a discovery tool.

A Summary of How Databases Are Organized on Library Websites

Table 20.8 shows you where the six main types of library databases are typically found on most library websites.

Table 20.8

Database	*Where to Find It*
A **catalog** for finding books and other media owned by a particular library	Catalogs are usually accessible from a campus library's homepage, but may not be explicitly labeled "catalog." You may need to ask library staff how to find and search it.
Article databases for finding articles (such as journal articles, conference papers, newspaper articles, and magazine articles)	These four types of databases are usually found by clicking a link labeled "Databases" (or something similar) on a campus library's homepage.
Content-specific databases for finding just one specific type of content—such as data, patents, videos, newspaper articles, etc.	
Publisher databases for finding sources produced by a particular publisher	
Pre-print databases for finding *manuscripts* of research articles and papers	
A **discovery tool** that can retrieve records from inside other databases	Libraries that offer a discovery tool will usually present them as a search box on the homepage that invites you to search for books, articles, media, and more.

Other Ways That Databases Differ

You've now learned that databases differ in the *types of sources they track*—for example, you'd use a catalog to find books and you'd use article databases to find articles. Furthermore, you may have noticed that some databases cover multiple subject disciplines, and we call them **multidisciplinary** databases. Examples of multidisciplinary article databases include Google Scholar, Web of Science, Scopus, and Academic Search.

There are also article databases that are **subject-specific**. For example, the article database PubMed is devoted to biomedical articles, the database Zoological Record provides articles on the topics of zoology and animal taxonomy, and the database Compendex is restricted to helping you find sources in the field of engineering.

But there's another way databases differ, and that's how far back in time they track sources. The article database PubMed covers sources from the 1960s to the present with some coverage going back to the 1940s, but nothing before that. However, in some disciplines, such as geology and zoology, older literature can still be important to refer to. So, the databases that serve those disciplines, such as GeoRef and Zoological Record, contain records for very old publications. Figure 20.7 shows a record from GeoRef for a source published in 1678.

1. An **account** of the **salt waters** of **Droytwich** in Worcestershire; sent by Dr. William Cole from Dr. Tho. Rastell, who hath lived many years upon the place, and hath there several phats of his own

Periodical

Cole, William In: Philosophical Transactions (1665), 1678, Vol. 12, pp.1059-1064.

Subjects: Hydrogeology; brackish **water**; brines; Droitwich England; England; Europe; Great Britain; **salt water**; springs; United Kingdom; Western Europe; Worcestershire England

Figure 20.7

Yet another way that databases differ is the search features they offer. This is an important topic that deserves your time and attention, so we'll cover it in Chapter 22.

You Should Also Be Aware of Reference Databases

There's one more type of database you should know about, but it's quite different from the databases we've discussed so far. **Reference databases** don't lead you to records for sources—rather, they contain information like you might find in a reference book.

For example, if you were to look up **water purification** in the reference database Wikipedia you would find an overview of this topic. Whereas if you searched the words **water purification** in an article database, you'd get a list of records leading to sources about that topic, like you saw in Figure 19.1.

The type of content available in reference databases is highly variable and there are lots of these databases available for free on the internet, such as:

- Wikipedia
- IMDb
- WolframAlpha

- Mindat.org
- GBIF
- NatureServe Explorer
- NIST Chemistry WebBook

Libraries also subscribe to certain reference databases, as you can see from the examples in Table 20.9.

Table 20.9

Database	*Subject Coverage*	*Source Types Tracked*
Acland's Video Atlas of Human Anatomy	Videos of human anatomy	(Not applicable)
Birds of the World	Species information birds	(Not applicable)
Global Plants	Images and information about plants	(Not applicable)
Reaxys	Chemical reaction and substance data	(Not applicable)
Statista	Business marketing data	(Not applicable)
Thermophysical Properties of Matter Database	Thermophysical properties	(Not applicable)
UpToDate	Guidance for clinicians in diagnosing illnesses	(Not applicable)
Current Protocols	Laboratory protocols	(Not applicable)

WHERE TO FIND REFERENCE DATABASES: You can usually find these databases by using the same link for finding article databases that's provided on your campus library's homepage. Once you click that link, search a for a specific database by title.

What about AI Literature Search Tools?

There are new tools on the horizon for searching sources, and they are driven by artificial intelligence (AI) technologies. Current options include Scite.ai, Elicit.org, and ResearchRabbit.ai. Because these tools are still in their infancy, some will likely go out of business and new ones will come into existence; eventually a few will corner the market and the options will stabilize. If you are interested in exploring what is currently available, just try the following Google search:

- **ai literature (search OR review) tools**

Be aware there are concerns about the use of these tools for searching sources, particularly if you want to conduct a comprehensive search. For most of these tools, you execute a search for sources by entering a few keywords, or you can enter a full sentence to state what you are looking for (this is called **natural language searching**). However, in most cases you know very little—or nothing—about how the AI tool is interpreting your search to find sources, or where it is looking for sources. When you don't know how a tool works, then you can't evaluate how well it did what you asked it to do. This is a problem with lots of AI technologies and we call it the **AI black box** problem.

Another concern is that these AI literature search tools might be using algorithms that prioritize the retrieval of sources that have been viewed or cited a lot, and this contributes to what's called the **Matthew effect**, in which certain sources that have become popular get even more popular as AI tools selectively retrieve them. It's basically an instance of the adage "the rich get richer and the poor get poorer," except in this case it applies to sources! To learn more about the black box problem or the Matthew effect, just search Google; there's plenty of interesting resources on both topics.

To sum up, what should science and engineering students know about databases?

1. Libraries provide access to a variety of types of databases.
2. A catalog is a database primarily used for finding books that a particular library owns, but you can also search a discovery tool to find books.
3. Article databases are the most abundant type of database and are used for finding articles and papers. Article databases can be multidisciplinary or subject-specific.
4. Some databases only track certain types of sources, such as patents, whereas other databases only contain sources published by a particular publisher.
5. Many libraries offer a discovery tool database, which is usually represented by a search box right on the library's homepage.
6. Databases differ in scope, coverage, time span, and search features.

What's Next?

In the next chapter we'll focus specifically on **database records** and the information they typically contain.

CHAPTER 21
DATABASE RECORDS

When you run a search in a database and get results, what exactly *are* those results? They're *records* that represent sources. If you have a basic understanding of records, then it'll be easier to grasp how database search features work and to leverage those features to your advantage. We'll cover database search features in the next chapter, but in this chapter you'll learn about database records and the information they contain. You'll also learn why you might find records for the same source in more than one database.

Brief Records and Full Records

When you search most databases, what you're really searching is a collection of **records**. Each record represents an individual source. The database will present your search results in the form of **brief records**, and if you click on a brief record, you'll pull up the **full record** for that source. (There are a few exceptions to this—for example, Google Scholar doesn't provide full records. Plus, there are a few databases that not only provide brief and full records for sources but also the full text of those sources.)

You're probably familiar with how brief and full records look when you search the commerce website, Amazon. If you run a search on "Jansport backpacks," you'll get a bunch of brief records for Jansport backpacks, and you can click on any of those brief records to see the full record for a particular backpack, with more detailed information. The same is true for bibliographic databases. When you run a search, your results will show brief records for individual sources, and when you click on any one of those brief records, you'll pull up the full record for that source, which will contain more detailed information.

You've seen brief records before in Figures 19.1 and 19.2. In Figure 21.1 you see some more brief records that resulted from a search for the keywords: **drones package delivery**. If you clicked on one of those brief records, you'd open up the full record which would have additional information, like you see in Figure 21.2.

You can see how full records contain all the bibliographic information you need to cite the source, plus they usually contain an abstract, as well as additional information about the source, such as subject terms, who funded the research, etc. But exactly what you see in a full record, and how it's organized and laid out, differs from database to database.

1. **Cost-effectiveness of drones for rural package delivery**
   Authors: Chalmers, L.C. et al.
   Source: *Journal of Drone Initiatives*, 2018, Vol. 23 (1) pages: 723-759
   View Abstract | Get full-text

2. **Maintaining drone fleets for package delivery: a checklist procedure**
   Authors: Chu, M. et al.
   Source: *23rd Conference on Unmanned Aerial Vehicle Strategies, 2020, pates: 45-69.*
   View Abstract | Get full-text

3. **Energy and emissions benefits for using drones as delivery vehicles**
   Authors: Rodrigues, L.S. et al.
   Source: *UAV Communications* 2022 Vol. 3 (8)
   View Abstract | Get full-text

Figure 21.1

Cost-effectiveness of drones for rural package delivery

Authors: Chalmers, L.C., Baker, M.S., Choi, P., and Slaughter, B.L.
Source: *Journal of Drone Initiatives*
Year: 2018
Vol. 23
Issue: 1
Pages: 723-759

Abstract: This research focuses on the benefits of using drones to deliver goods to rural residences in place of vehicular transport. We compare costs for labor, fuel, and maintenance. We also explore benefits such as reduction in greenhouse gas emissions. We outline start-up costs and when pay-back will be achieved.

Subject terms: drones, package delivery, economics, efficiency, energy consumption trade-offs, greenhouse gas emissions

Major research areas: Engineering, Transportation, Computer Science

Language: English
Document type: Review article
Funding sources: U.S. Department of Energy
Corresponding author contact: University of Springfield
Journal ISSN: 8763-1245
Accession number: 0098767

Figure 21.2

Note that a full record is sometimes referred to as a **bibliographic record**. This is a sensible name because full records contain bibliographic data. Both full and brief records are also sometimes called **citations**. Figure 21.3 shows the search box for the article database PubMed along with a description of this database. It's described as comprising "more than 36 million citations for biomedical literature from MEDLINE, life science journals, and online books." In that description, you could exchange the word "citations" for "bibliographic records" or just "records," because in this context they all mean the same thing.

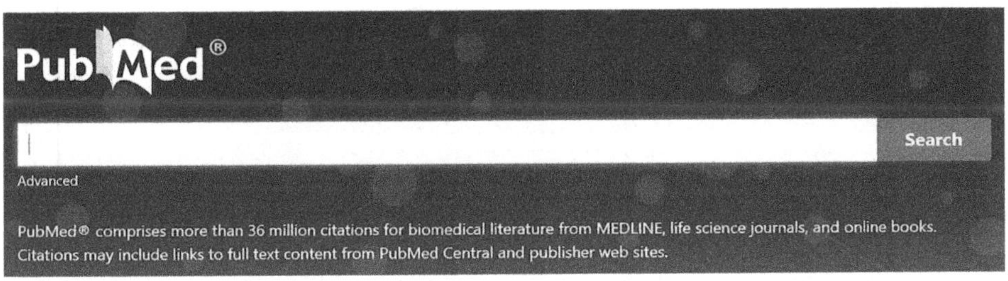

Figure 21.3

Database Records Usually Indicate How to Get to the Actual Source

Database records usually provide some way of leading you to the actual source. Take, for example, the two book records shown in Figure 3.5. The first record provides a link to view the full text of the book online. The second record provides a call number and location for finding the print book on the shelves in the library's collection.

Sometimes when you click a link within a record to get to the full text of the source, you'll encounter an intermediate page where you have to look for another link to click to get through to the full text. Alternatively, that intermediate page may instead offer an option to order the source through your library's **document delivery/interlibrary loan service**. For more information about how to get to the full text of a source and the problems you might encounter, see Chapter 28.

Databases Only Contain Records for Sources That Meet Their Inclusion Criteria

When a new source is published, a record for that source will be added to any database for which that new source meets the database's inclusion criteria. Recall from the last chapter that databases differ in a variety of ways including the types of sources they track (books, journal articles, patents, etc.) as well as the subject matter of those sources. **Inclusion criteria** are the guidelines that spell out those parameters for each database. For example, the article database Web of Science only collects records for articles from scholarly journals that meet certain guidelines for quality, and here's an excerpt from those guidelines:

> The basic principles of our selection process remain the same: objectivity, selectivity and collection dynamics. We use a single set of 28 criteria to evaluate journals; these are divided into 24 quality criteria designed to select for editorial rigor and best practice at the journal level, and four impact criteria designed to select the most impactful journals in their respective fields using citation activity as the primary indicator of impact.[1]

Figure 21.4 shows a source that met Web of Science's inclusion criteria and was added to the Web of Science database.

Anatomical, morphometric, and stratigraphic analyses of theropod biodiversity in the Upper Cretaceous (Campanian) Dinosaur Park Formation

Cullen, TM; Zanno, L; (...); Evans, DC

Sep 2021 | CANADIAN JOURNAL OF EARTH SCIENCES 58 (9), pp.870-884

The Dinosaur Park Formation (DPF) of Alberta, Canada, has produced one of the most diverse dinosaur faunas, with the record favouring large-bodied taxa, in terms of number and completeness of skeletons. Although small theropods are well documented in the assemblag ... Show more

11 Citations

107 References

Figure 21.4

However, this source also met the inclusion criteria for EBSCO's database Academic Search, and here's the brief record (Figure 21.5) for it in *that* database.

1. **Anatomical**, **morphometric**, and **stratigraphic analyses** of **theropod biodiversity** in the **Upper Cretaceous** (**Campanian**) **Dinosaur Park Formation**[1].

Academic Journal

By: Cullen, Thomas M.; Zanno, Lindsay; Larson, Derek W.; Todd, Erinn; Currie, Philip J.; Evans, David C. Canadian Journal of Earth Sciences. 2021, Vol. 58 Issue 9, p870-884. 15p. DOI: 10.1139/cjes-2020-0145.

Subjects: SAURISCHIA; **DINOSAURS**; FRONTAL bone; **BIODIVERSITY**; FOSSILS; **PARKS**; NORTH America; ALBERTA; Nature **Parks** and Other Similar Institutions

PDF Full Text (8.8MB)

Figure 21.5

---

[1]To read more details about the **Web of Science Journal Evaluation Process and Selection Criteria**, see: https://clarivate.com/products/scientific-and-academic-research/research-discovery-and-workflow-solutions/web-of-science/core-collection/editorial-selection-process/editorial-selection-process/

So, the same source might fit the inclusion criteria for more than one database, which means there can be some overlap in records you find when you search different databases.

However, not all records overlap! EBSCO's Academic Search database collects records for certain magazine articles that fit its inclusion criteria, like the one shown in Figure 21.6.

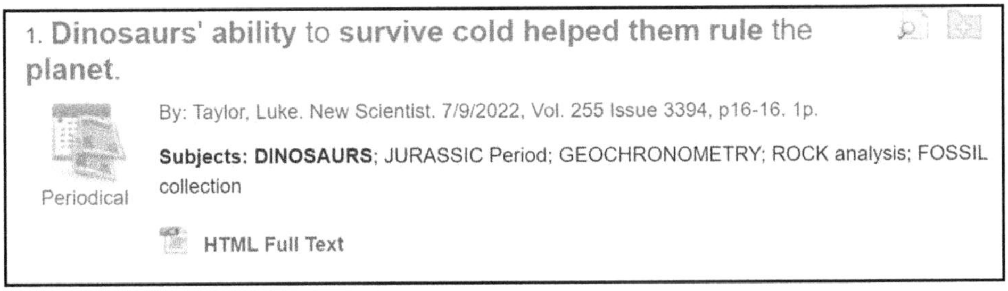
1. Dinosaurs' ability to survive cold helped them rule the planet.

Periodical

By: Taylor, Luke. New Scientist. 7/9/2022, Vol. 255 Issue 3394, p16-16. 1p.

Subjects: DINOSAURS; JURASSIC Period; GEOCHRONOMETRY; ROCK analysis; FOSSIL collection

HTML Full Text

Figure 21.6

But, you would *not* find that record in Web of Science, because Web of Science doesn't include *any* records for magazine articles. Meanwhile, the database Compendex has inclusion criteria stipulating it will only collect records for sources about engineering, while Zoological Record is a database that only collects records for sources about zoology. There probably aren't *any* overlapping records in these two databases because they cover such different subject disciplines.

To sum up, what should science and engineering students know about database records?

1. When you search most databases and get results, your results represent *records* for sources presented in a brief format. You can click on the brief record to open it and view the full record, which contains more details.
2. Database records usually provide some way of getting to the full text of the source.
3. You may find some of the same records in different databases, if those databases have overlapping inclusion criteria.

What's Next?

In the next chapter we'll focus on **database search features**. Understanding search features will help you become a much more efficient and effective searcher!

CHAPTER 22
DATABASE SEARCH FEATURES

Search features are what make databases cool and fun to use. Seriously—it's true! In this chapter you'll learn about different search features that databases offer and how they work. You'll also learn that some databases are specifically designed with search features to help science and engineering researchers find the information they need more easily.

What are Search Features?

Search features refer to the different options offered by a database for constructing your search and refining and sorting your results. One type of search feature is the ability to search your keywords in a specific part of a database record, such as the source's title, the abstract, or subject terms. Other search features allow you to refine your results in various ways, such as by limiting to English-language sources only, or to certain types of sources like review articles, or to sources published over a certain time span, such as the last five years. Yet another type of search feature allows you to sort your results in a variety of ways such as by relevance, date published, or by the number of times sources have been cited.

To Understand Search Features You Must First Understand Full Records

Many of the search features offered by a database directly relate to the data that are included in a full record (a.k.a. the bibliographic record). You were introduced to full records in the last chapter, but now let's talk about them in more detail to show how they relate to search features. We'll start at the beginning, when a full record gets added to a database, and we'll use an imaginary example.

Suppose a database company called "HiQual Databases" decides there's a good market for developing and licensing an article database that tracks journal articles about planetary moons. The company moves forward with their plan, and decides to call their new database MoonCites. HiQual employees on the MoonCites production team would set up inclusion criteria for MoonCites and then gather bibliographic data for already-published journal articles about moons. Then, they'd create full records for all the articles that fit the inclusion criteria. (Refer to the previous chapter for more information about inclusion criteria.)

They'd also start monitoring *newly* published journal articles on the topic of moons, and add full records for those articles as they were published. So, MoonCites, like all article databases, would be continuously growing, with new records added as new articles are published that fit the inclusion criteria.

Let's look at how a full record would be added to MoonCites, using the article below as an example:

> Kanumba, A. & Smith, T. (2022). The origin and formation of Titan, Saturn's largest moon—a comprehensive review. *Journal of Moon Research 26*(1): 34–56.

In Figure 22.1 you see data fields such as author, article title, journal title, etc., that must be populated to create a full record for this article. Populating those data fields might be done through an automated process or it might be done manually by a HiQual employee, or using a combination of automated and manual methods.

As you learned in the last chapter, full records contain all the relevant **bibliographic details** about the source (author, article title, etc.), plus they usually provide an **abstract** (which is a short synopsis of the contents of the source). Full records for the MoonCites database also contain information about the language the article is written in, the document type, the planet the moon is associated with, the moon's size, as well as additional subject terms that describe the contents of the article. **Subject terms** are keywords that have been chosen to reflect what the article is about. Subject terms are also sometimes called **subject headings**, **subject descriptors**, **index terms**, or just **descriptors**.

So, Figure 22.1 shows a typical-looking full record for an article database. In the next section you'll learn how elements in this full record drive what search features a database can offer.

Author: Kanumba, Akinyi; Smith, Tyrone
Article title: The origin and formation of Titan, Saturn's largest moon – a comprehensive review
Journal title: Journal of Moon Research
Publication year: 2022
Volume: 26
Issue: 1
Pages: 34-56
Abstract: We review the literature on the origin and formation of the moon Titan including its geologic history, composition, and atmosphere. All the latest literature and findings about Titan's surface features are summarized. We further examine conditions conducive to life, such as the formation of complex molecules and likely surface habitats. Controversies and avenues for future research are addressed.

Language: English
Document type: Review article
Moon size: >1000-mile radius
Associated planet: Saturn
Subject terms:

- Titan
- Origin
- Co-accretion
- Atmospheric evolution
- Surface evolution
- Surface features
  - Lakes
  - Craters
  - Volcanoes
  - Mountains
- Composition
  - Ice
  - Rock
- Prebiotic conditions

Figure 22.1

How Search Features Relate to Full Records

In Figure 22.2, you see the search interface for the database MoonCites showing some of this database's search features. (A **search interface** is the user-friendly screen where you enter your keywords.) Notice your keywords will be automatically searched as "Topic" words. However, you could change the drop-down selection and search in particular fields of a record such as article title, abstract, author, subject terms, etc.

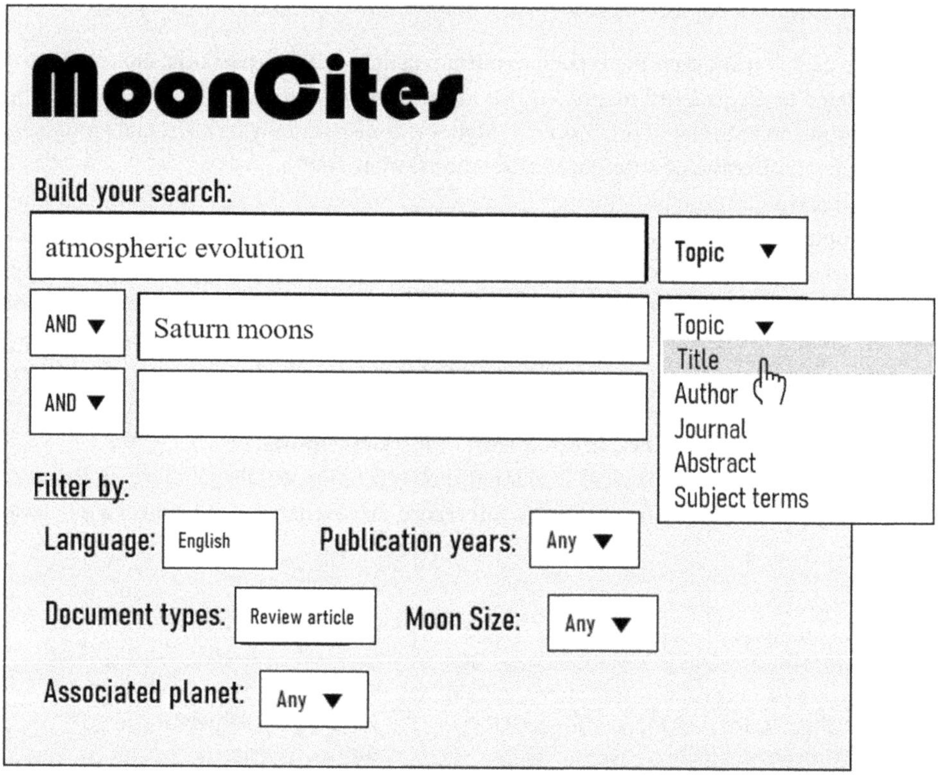

Figure 22.2

Below the boxes where you enter your keywords, you have the ability to refine your search by language, publication year, document type, moon size, and associated planet. These refinements are called different things in different databases. As you can see from Figure 22.2, in MoonCites they're called **filters**. In other databases they might be called **limits**, **refinements**, **search options**, or something similar.

Now suppose you're searching MoonCites for articles about the atmospheric evolution of Saturn's moons and you're specifically looking for review articles written in English. You could set up the search like you see in Figure 22.2, where you search "Saturn" and "moons" in the title field, while you search for your other terms (atmospheric evolution) in the default "Topic" field. Then, you'd set your language filter for "English" and your document type filter for "Review article."

When the database processes this search, the records retrieved must have *all* the keywords you entered and two of those keywords ("Saturn" and "moon") *must* be in the article title. The records retrieved must also fit the filters you stipulated ("English-language" and "review articles"). Figure 22.3 shows the first five records that were retrieved. Note that this database retrieves the plural, singular, and possessive forms of your keywords, and it highlights them.

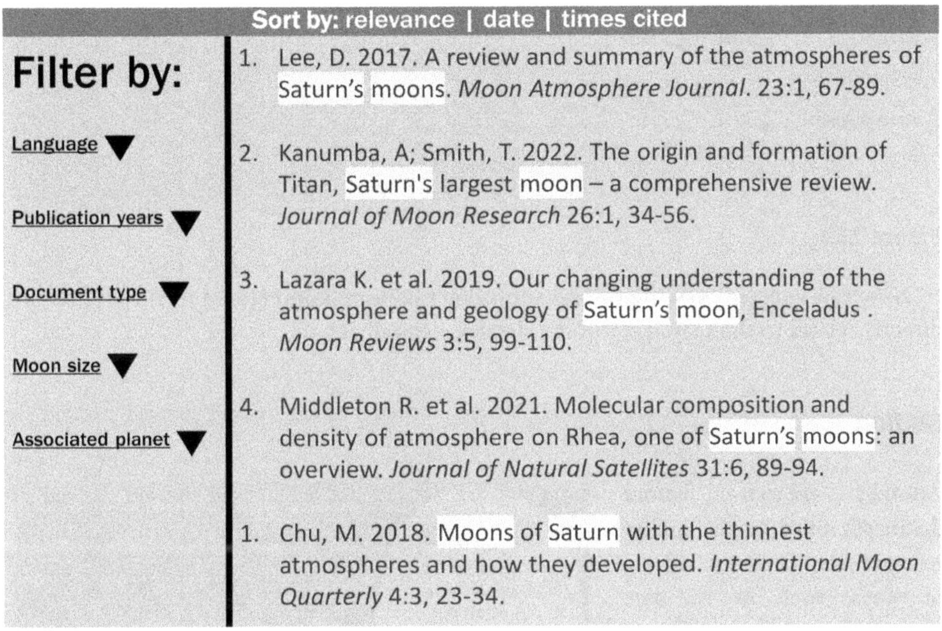

Figure 22.3

But how did the database know that these records were appropriate to retrieve? And specifically, how did the database know these records represent review articles published in English? Because it found that information in the **full record**. In other words, the database looks in its full records for matches to the keywords you entered, and it also checks full records for any refinements you requested, such as English-language only. It then delivers *only* those results that match, and it delivers them in the form of brief records, like you see above.

Using that second record by A. Kanumba as an example, when you click on it to pull up the full record, you can see that the database found a match for everything you told it to find. It found matches for "Saturn" and "moons" in the title, and it found a match for "atmospheric" and "evolution" in the subject terms. Plus, the full record indicates that the article is in "English" and is a "Review article" (see Figure 22.4).

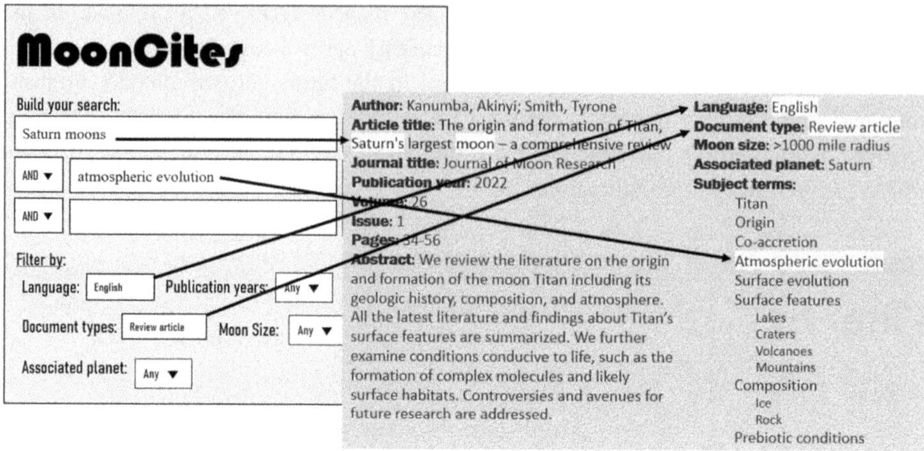

Figure 22.4

Now can you see how the information that's collected and stored in the full record directly relates to the search features a database can offer?

Sorting

Another search feature databases offer is the ability to sort results in a variety of ways, such as by date published, relevance, times cited, etc. Sorting options usually appear at the top of your list of results, like you see in Figure 22.3.

Most databases default to sorting your results by **relevance** (you'll learn more about relevancy sorting in Chapter 26). But if you prefer, you can re-sort them by publication date to see the most recently published articles first, or you can sort your results by times cited. **Times cited** refers to the number of times other sources have cited a particular source.

If you're using a database that allows you to sort in this way, then you can sort your results to see the most highly cited articles at the top.

When articles are highly cited, that's usually an indication they contain information that lots of other researchers found useful. However, an article can get cited a lot for other reasons as well—for example, it might contain information that's controversial or contentious. For more information about article citation metrics, see Appendix B.

Databases with Information-Rich Records Will Offer Excellent Search Capabilities

In general, the more information-rich the full records are for a database, the more search features the database will be able to offer users. A database with information-rich records will also typically be more expensive to produce, because creating detailed, accurate, full records usually requires some human intervention—and the humans doing this work need to be highly qualified.

Figure 22.5 shows the types of data you'd find in a full record from PubMed (those data are consolidated in this image to fit in as much as possible).

Article title: Use of anastrozole for breast cancer prevention (IBIS-II): long-term results of a randomized controlled trial
Authors: Jack Cuzick, Ivana Sestak, John F Forbes, Mitch Dowsett, Simon Cawthorn, Robert E Mansel, Sibylle Loibl, Bernardo Bonanni, D Gareth Evans, Anthony Howell.
Journal: Lancet. **Year:** 2020 **Volume:** 395 **Issue:** 10218 **Pages:** 117-122
doi: 10.1016/S0140-6736(21)00427-X. **PMID**: 33640066
Abstract:
Background: Two large clinical trials have shown a reduced rate of breast cancer development in high-risk women in the initial 5 years of follow-up after use of aromatase inhibitors (MAP.3 and International Breast Cancer Intervention Study II [IBIS-II]). Here, we report blinded long-term follow-up results for the IBIS-II trial, which compared anastrozole with placebo, with the objective of determining the efficacy of anastrozole for preventing breast cancer (both invasive and ductal carcinoma in situ) in the post-treatment period.
Methods: IBIS-II is an international, randomised, double-blind, placebo-controlled trial. Postmenopausal women at increased risk of developing breast cancer were recruited and were randomly assigned (1:1) to either anastrozole (1 mg per day, oral) or matching placebo daily for 5 years. After treatment completion, women were followed on a yearly basis to collect data on breast cancer incidence, death, other cancers, and major adverse events (cardiovascular events and fractures). The primary outcome was all breast cancer.
Findings: 3864 women were recruited between Feb 2, 2003, and Jan 31, 2012. 1920 women were randomly assigned to 5 years anastrozole and 1944 to placebo. After a median follow-up of 131 months (IQR 105-156), a 49% reduction in breast cancer was observed for anastrozole (85 vs 165 cases, hazard ratio [HR] 0·51, 95% CI 0·39-0·66, p<0·0001). The reduction was larger in the first 5 years (35 vs 89, 0·39, 0·27-0·58, p<0·0001), but still significant after 5 years (50 vs 76 new cases, 0·64, 0·45-0·91, p=0·014), and not significantly different from the first 5 years (p=0·087). Invasive oestrogen receptor-positive breast cancer was reduced by 54% (HR 0·46, 95% CI 0·33-0·65, p<0·0001), with a continued significant effect in the period after treatment. A 59% reduction in ductal carcinoma in situ was observed (0·41, 0·22-0·79, p=0·0081), especially in participants known to be oestrogen receptor-positive (0·22, 0·78-0·65, p<0·0001). No significant difference in deaths was observed overall (69 vs 70, HR 0·96, 95% CI 0·69-1·34, p=0·82) or for breast cancer (two anastrozole vs three placebo). A significant decrease in non-breast cancers was observed for anastrozole (147 vs 200, odds ratio 0·72, 95% CI 0·57-0·91, p=0·0042), owing primarily to non-melanoma skin cancer. No excess of fractures or cardiovascular disease was observed.
Interpretation: This analysis has identified a significant continuing reduction in breast cancer with anastrozole in the post-treatment follow-up period, with no evidence of new late side-effects. Further follow-up is needed to assess the effect on breast cancer mortality.

Publication types
- Comparative Study
- Randomized Controlled Trial
- Research Support, Non-U.S. Gov't

MeSH terms
- Administration, Oral
- Adult
- Aged
- Anastrozole / adverse effects
- Anastrozole / therapeutic use*
- Aromatase Inhibitors / therapeutic use*
- Breast Neoplasms / drug therapy*
- Breast Neoplasms / epidemiology
- Breast Neoplasms / mortality
- Breast Neoplasms / prevention & control*
- Carcinoma, Ductal, Breast / drug therapy
- Carcinoma, Ductal, Breast / pathology
- Carcinoma, Intraductal, Noninfiltrating / drug therapy
- Carcinoma, Intraductal, Noninfiltrating / pathology
- Female
- Follow-Up Studies
- Humans
- Incidence
- Middle Aged
- Placebos
- Treatment Outcome
- United Kingdom / epidemiology

Substances
- Aromatase Inhibitors
- Placebos
- Anastrozole

Figure 22.5

Notice this record contains all the bibliographic information for the source (title, author, journal, etc.), an abstract, as well as the publication type of the source, which—in this database—means the *type of study* or *study design*. In this example, the source is both a "comparative study" as well as a "randomized controlled clinical trial."

You also see a detailed list of **Medical Subject Headings** (**MeSH terms**), as well as the substances discussed in the article. While you don't see language listed, it's coded into this record but just not visible. Notice that those MeSH terms capture a lot of information about this article, including all of the following details and more: age group, sex, species, drug, disease, etc.

The amount of information provided in this record gives you an idea of how sophisticated the PubMed search interface must be. And it is! In PubMed, you can filter searches by article/publication type, which provides dozens of options to select from, as well as by species, language, sex, journal, and age group. Take a look at PubMed's options for filtering your results—it's rather amazing (see Figure 22.6).

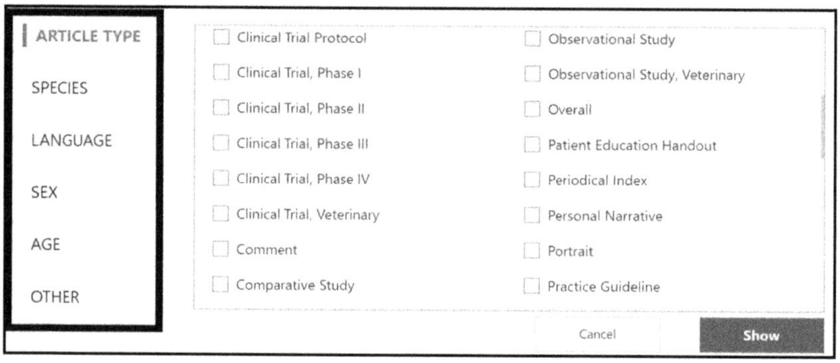

Figure 22.6

Can you see how these filters are designed with the particular needs of biomedical researchers in mind? Suppose a team of biomedical researchers was working on a project where they needed to collect articles on the use of the drug anastrozole for breast cancer, but they only want to see articles that are randomized controlled trials of human females who are middle-aged or older. They could enter the keywords "breast cancer anastrozole" into PubMed, then set their filters for the appropriate article/publication type, species, sex, and age. This would allow them to quickly find the best, most relevant sources. In fact, the article record from PubMed shown in Figure 22.5 would be one of the articles their search would retrieve.

Meanwhile, SciFinder is a database designed with search features that are especially useful to chemists. In the SciFinder search interface you can search for articles, but you can also search for substances, reactions, suppliers, etc. And, when you run a search, you're offered lots of filters that are useful to chemists; for example, you can refine your search to just those articles that are talking about a substance in a particular role, such as how to prepare it, how to use it, etc.

Now compare the search features offered in PubMed and SciFinder with those offered by Google Scholar (see Figure 22.7)—which only provides a way to refine your results by publication date, and to very limited document types (just review articles, in fact).

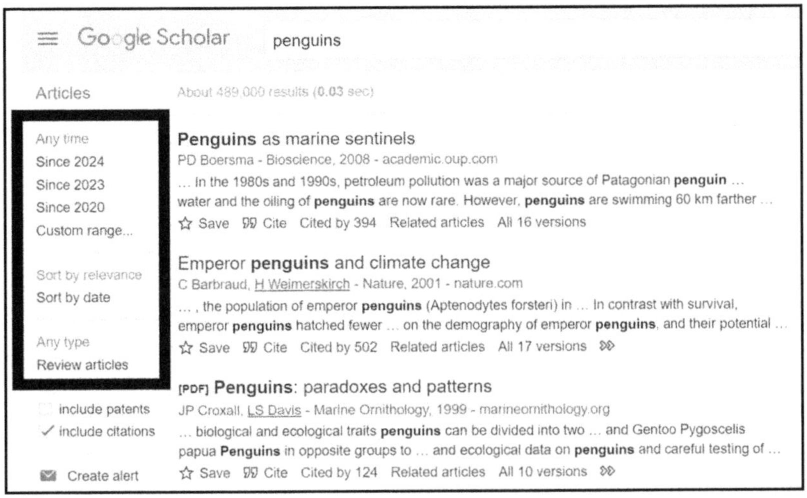

Figure 22.7

As you can see, Google Scholar offers almost no search features and certainly none that are designed to help researchers in a particular discipline find information. So, what can you conclude from all this? A database that's designed for finding information in *your specific discipline* may help you find relevant sources faster and with less fuss.

To sum up, what should science and engineering students know about database search features?

1. Many article databases have search features that allow you to search for sources more efficiently and effectively.
2. Search features include the ability to search your keywords in a specific part of a database record, to refine your results in various ways, and to sort your results in various ways.
3. Some databases offer search features that are designed to help researchers *in a specific discipline* zero in on useful sources.

What's Next?

Now that you have a much better understanding of databases and their search features, let's move on and talk about how to **choose the best databases** to search.

CHAPTER 23
CHOOSING THE BEST DATABASES

By now you've learned a lot about databases—but how do you choose the best ones to search? This chapter will explain! You'll learn how to determine which databases are available to you, and which to search based on the characteristics and scope of your topic. This chapter will also offer advice on how many databases you should search.

Database Availability

When choosing the best databases to search, the first and most obvious constraint you'll face is *availability*. You'll only be able to search databases that are available to you, which could include some or all of the following:

1. **Campus library databases**
2. **Public library databases**
3. **State library databases**
4. **Employer databases**
5. **Free databases**

Let's look at each of these options in more detail and discuss what you need to know about availability.

1 Campus Library Databases

While you're a student enrolled at a college or university, you'll be able to search all the databases your campus library provides, and most campus libraries offer quite a few. However, different campus libraries make different choices about which databases to offer their students and faculty. For example, a campus library at a university that has an engineering program would probably license the engineering-specific article database Compendex. But libraries on campuses that don't have an engineering program wouldn't bother to spend money licensing this database, because their students and faculty don't need it. Also, campus libraries with bigger budgets can afford more databases than campus libraries with smaller budgets.

Just so you know, when campus libraries sign license agreements to acquire a database, the license requires the library to restrict access to students, faculty, and staff who are currently affiliated with that campus. How do campus libraries comply with

this requirement? The library configures their databases so they're only accessible to users who are on that campus's computer network or on the virtual private network (VPN) for that campus. Anyone attempting to access library databases from outside that network will receive a prompt to authenticate using their institutional email username and password. After they enter their credentials, they'll be given access to the database. Obviously, anyone *without* an email account at that institution will not be able to log in, and so will be denied access.

Something else to keep in mind is that when you graduate and no longer have an affiliation with a campus, you'll lose access to most of the databases your campus library offers. You'll probably still be able to access the library's catalog, discovery tool, or any free databases your library presents on their website like Google Scholar or PubMed, but not subscription databases like GeoRef, Web of Science, Scopus, etc.

However, there is a way around this. If you happen to live near a campus library and that library allows members of the public to walk in and use library resources, then you may be able to get on the campus network from inside the library. That will allow you to access that library's databases and other electronic collections. Just call your local campus library to see what sort of community access they offer.

2 Public Library Databases

Most public libraries offer a catalog and/or discovery tool you can search as well as a selection of article databases. However, the databases public libraries offer are usually geared toward helping you find popular sources rather than scholarly sources. This makes sense because most public library users are looking for information that's written for a general audience and is practical and/or entertaining—not scholarly. To find out which databases your public library offers, just visit your public library's website, or stop by your public library in person.

3 State Library Databases

In the United States, some states license article databases as well as a few other types of databases, and anyone living in the state can access these databases. To find out what's available to you, do a Google search for your state library (assuming there is one!) and then peruse their website to see what databases are available to access (if any).

4 Employer Databases

Some companies license databases and those that do are usually engaged in some sort of research, such as pharmaceutical companies, certain types of engineering companies, etc. If you're working for a company, you'll likely be told when you are hired what databases the company offers and how to access them.

5 Free Databases

Some databases are free for anyone to search at any time. In fact, remember that list of database types you saw at the beginning of Chapter 20? Here is it again:

1. A **catalog** for finding books and other media owned by a particular library
2. **Article databases** for finding articles (such as journal articles, conference papers, newspaper articles, and magazine articles)
3. **Content-specific databases** for finding one specific type of content, such as data, patents, videos, newspaper articles, etc.
4. **Publisher databases** for finding sources produced by a particular publisher
5. **Pre-print databases** for finding manuscripts of research articles and papers
6. A **discovery tool** that can retrieve records from inside other databases.

To see which of those types of databases are free to search, consult Table 23.1. Also, don't forget in Chapter 20 you learned about **reference databases** and you saw some examples of those that are freely available.

By the way, a database is considered free when anyone can search it and pull up records for sources. However, that doesn't mean you'll be able to access the full text of those sources. So *searching* is free, but obtaining the full text of the sources you retrieve may not be free. You may be asked to authenticate (log in) or pay to view the full text.

Table 23.1

Database Type	*What You Can Search For Free*
Catalogs	Library catalogs are usually free to search. Just go to any library's website and if you see an option to search the catalog, you should be able to search it to see what books that library has in their collection. Also, don't forget you can search for books in WorldCat, Google Books, Amazon, etc. These resources are discussed in more detail in Chapter 3.
Article databases	Most article databases are *not* free to search, but there are a few that are, such as: Agricola, ADS (Astrophysics Data System), ERIC, Google Scholar, PubMed, etc. If you are looking for additional free databases to search, then run these Google searches to find more: • **free article databases** • **free article databases [plus your disciplinary area]** • **free databases [plus your disciplinary area]**
Content-specific databases	Many content-specific databases are *not* free to search, but some are. Consult Chapters 10, 12, and 13 for options.
Publisher databases	Most publisher databases are free to search, such as ScienceDirect, Wiley Online Library, etc. You don't necessarily have to access them from a library website, just pull them up in a Google search.
Pre-print databases	All pre-print databases are free to search. To find them, consult: • https://en.wikipedia.org/wiki/List_of_pre-print_repositories • https://doapr.coar-repositories.org/repositories/
Discovery tools	Many library discovery tools are free to search. Just go to any campus library website and if you see a discovery tool, then simply run a search.

If that happens, try searching the title of the source in Google to see if you can find a free copy somewhere. If not, and if you're not affiliated with a campus library that can provide the source through interlibrary loan/document delivery, then you may be able to use your public library's interlibrary loan/document delivery service instead. However, first check whether your public library charges fees for using this service or puts limits on the number of times you can use it.

Choose Databases Based on Your Information Need

Once you've determined which databases are available to you, you'll need to decide which to search. There may be several dimensions to your particular information need that will dictate the types of databases you should search. One dimension to consider is your assignment's source requirements. If you're working on an assignment that requires you to find certain *types* of sources, then you'll need to search the databases that contain records for those types of sources. Other dimensions include the *subject matter* of your information need, the *scope* of your need, and whether you need information with certain types of *qualities*. Let's explore these dimensions in more detail to see how they relate to choosing databases.

Assignment Source Requirements

If you're working on an assignment that requires you to find particular source types—such as books, journal articles, patents, etc.—then you'll need to find and use the databases appropriate for those source types. For example, use your library's **catalog** to find books, use **article databases** to find journal articles and conference papers, and use **content-specific databases** to find patents, data, etc. Or, if your library offers a **discovery tool** for finding books, articles, media, and more, then you can search that and then refine your results to the particular source type you need—as you've learned to do in previous chapters. Refer to Chapter 18 for more information about assignment source requirements and refer to Chapter 20 for more guidance on which databases to use for finding particular source types.

Subject Matter

Subject-specific databases can be especially useful to search, as you learned in the last chapter. To find the databases recommended for a particular subject area, go to your campus library's website and look for a link to "databases"—or look for any link with a similar name. Once you click it, look for ways to view your library's databases by **subject** or **discipline**, like you see in Figure 23.1. If you don't see a link to get to databases on your library's homepage, or you don't see a way to view databases by subject, then ask for help.

Figure 23.1

Let's suppose you used the drop-down selection (shown in Figure 23.1) to view engineering databases. Typically, you'll land on a page with a list of databases recommended for engineering topics, and it might include a mix of article databases (both subject-specific and multidisciplinary), content-specific databases, publisher databases, and reference databases, like you see in Figure 23.2.

The database descriptions shown in Figure 23.2 are clear and concise, but this is not always the case. Some libraries provide rather confusing database descriptions. If that's the case at your library, then ask for clarification.

Note that even if you are no longer affiliated with a campus library, you can still go to any campus library website and seek out the databases they recommend for your subject area. You can even try clicking the links to see if you can get access, as a few of them may be free to search.

ENGINEERING DATABASES

ASTM Compass
Content-specific database for finding engineering standards.

Compendex
Subject-specific article database for engineering.

Derwent Innovations Index
Content-specific database for finding patents.

Google Patents
Content-specific database for finding patents.

IEEE Explore Digital Library
Publisher database for the Institute of Electrical and Electronics Engineers.

Knovel Critical Tables
Reference database of properties of commonly used chemical compounds.

Scopus
Multidisciplinary article database.

Web of Science
Multidisciplinary article database.

Figure 23.2

Scope

To choose databases based on the scope of your information need, use the list below. If you need more information about the databases mentioned below, refer to Chapter 20.

- If you need discrete facts then try a Google search, or consult reference databases like those you learned about at the end of Chapter 20. Or, you can search your library's **catalog** (or **discovery tool**) to find reference books that might have the information you need.
- If you need a quick overview of a topic, then try a Google search or an online encyclopedia like Wikipedia, or a **reference database** provided by your library such as Credo Reference or Gale Virtual Reference Library.
- If you need a comprehensive, broad overview of a topic, then look for a book about that topic using your library's **catalog** or **discovery tool**. You can also check Amazon, Google Books, and/or WorldCat.
- If you need detailed research information, then use **article databases** like Google Scholar, PubMed, IEEE Xplore, etc., to look for journal articles and conference papers etc. Or, you could try using your library's **discovery tool**, and then limit your results to journal articles and/or conference papers.

Qualities

If you need information that has certain types of *qualities*, then use the list below to choose the best databases—and again, consult Chapter 20 for specific details about the databases listed here.

- If you need scholarly information, then use your library's **catalog** to find scholarly books and/or use **article databases** to find journal articles and conference papers. You can also try searching your library's **discovery tool** for these types of sources.

- If you need popular information, then use Google to find websites, or use your library's **discovery tool**, **article databases**, or any **content-specific databases** that provide access to newspaper and magazine articles. For popular books, consult your campus or public library's **catalog**, or try searching an online bookseller like Amazon.
- If you need science or engineering research information that's original and unprocessed, that is to say **primary information**, then consult sources such as research articles or conference papers. To find those types of sources, use **article databases** and/or your library's **discovery tool**.
- If you need summaries, analyses, or interpretations of science or engineering information, that is the say **secondary information**, then consult sources such as books or review articles. Use the **catalog** to find books and use **article databases** to find review articles. Or, use your library's **discovery tool** to look for either.

If you are stumped about which databases to search, then ask library staff for advice. Your campus library may have a librarian who specializes in the subject area that you are researching, and can quickly tell you the best databases to search for your topic. Alternatively, you can try using **Generative Artificial Intelligence (AI)** services like **ChatGPT** (https://chat.openai.com/) to get ideas for which databases to search. Just provide a prompt to indicate what you are researching and then ask for recommended databases to search. Sometimes the advice is great, other times it's terrible, but it's worth a try and the advice these AI services provide may get better over time. For more information about Generative AI and for ideas for other services to query (like Bing Chat, Bard, etc.) consult the Wikipedia page on Generative AI: https://en.wikipedia.org/wiki/Generative_artificial_intelligence

Article Databases vs. Publisher Databases

For a refresher on what article databases are and what publisher databases are, see Chapter 20. If you're looking for articles on a topic, as a general rule it's best to stick with searching **article databases**, because those databases track articles *across multiple publishers*. After all, why restrict your search to the offerings of just one publisher? However, the situation is a bit different in certain disciplines like engineering, computer science, and math. In those disciplines, **publisher databases** can be essential to search for two reasons. The first reason is that not all libraries offer good article databases devoted to those disciplines, and so publisher databases can fill that gap. The second reason is that in some disciplines, there's just one publisher that publishes most of the sources in that discipline, and so searching that publisher's database will result in a pretty comprehensive search. A good example is the database IEEE Xplore, which tracks sources published by IEEE (Institute of Electrical and Electronics Engineers). IEEE is the main publisher of sources in electrical engineering, so if you search IEEE Xplore for an electrical engineering topic, your search should be fairly comprehensive.

Should You Use Pre-Print Databases?

You may be wondering whether you should refer to and cite manuscript versions of journal articles and papers that you find in pre-print databases (these databases are explained in Chapter 20). Certainly post-prints are safer to cite than pre-prints, because they've undergone peer review. But how you should use these databases and their contents may depend on the discipline you're in. So, get advice from your instructors, or simply find and use the final, published version of the journal article or conference paper instead of the pre-print version.

How Many Databases Should You Search?

The number of databases you should search will depend on the nature of your topic and how thorough you need to be. If you're working on a course assignment and you just need to find a few relevant sources, then searching your library's **discovery tool** might be all you need. However, if you're working on a more serious research project (either while in college, or graduate school, or in your career) you may need to do a comprehensive search to find everything on your topic that's relevant. Or, if you're compiling a review article then you'll want to conduct an especially comprehensive search (consult Chapter 8 for more information about review articles.) In these situations, searching your library's discovery tool—and nothing else—would be completely insufficient.

Let's suppose you're researching the moisture-retaining properties of biological soil crusts and need to conduct a comprehensive search for journal articles on that topic. In that case you should search any article database you have access to that's likely to have journal articles on this topic—including article databases that are multidisciplinary as well as subject-specific.

This topic is relevant to multiple disciplines (such as biology, forestry, and environmental science) so you'll want to select relevant databases from any of those disciplines. Assuming you have access to databases from a campus library, you might want to explore which databases are available in all those subject areas, and then make a list of the databases that look most promising to search. Below you see a list of databases that were chosen for this topic:

1. **Agricultural and Environmental Science Collection**
   Subject-specific article database for agriculture and environmental science.
2. **BIOSIS**
   Subject-specific article database for biology.
3. **CAB Abstracts**
   Subject-specific article database for applied sciences like agriculture, forestry, environmental science, etc.
4. **GreenFILE**
   Subject-specific article database for environmental science.
5. **Web of Science**
   Multidisciplinary article database.

Or, suppose you need to find journal articles about the knowledge, attitudes, and practices of U.S.-based obstetricians regarding pregnancy in women with disabilities. For this topic, you could probably do a very comprehensive search in the four databases shown below, but if you wanted to be extra thorough you could also search a couple of multidisciplinary article databases like Web of Science and Scopus.

1. **CINAHL**
   Subject-specific article database for nursing and allied health.
2. **PsycINFO**
   Subject-specific article database for psychology.
3. **SocINDEX**
   Subject-specific article database for sociology.
4. **PubMed**
   Subject-specific article database for biomedicine and clinical medicine.

Let's look at one more example. Suppose you were researching how to engineer buildings to withstand earthquakes. For this topic you could probably do a comprehensive search just using the engineering database Compendex, although you might also try searching multidisciplinary databases like Web of Science and Scopus.

So, when you need to do a comprehensive search, the number of databases you should search really depends on the nature of your topic—and particularly whether your topic is multidisciplinary and likely to be covered in more than one database. And obviously you'll only be able to search the databases available to you.

Get Help!

Understanding which databases are available to you, which would be best to search, and how many to search is not easy because each research topic is different. So, save yourself a lot of time and ask for help. If you are affiliated with a campus library, then contact your library (look for a chat service right on your library's homepage) and ask to be put in touch with a librarian who can recommend which databases to search. For more information about how librarians can help with your research, see Appendix A.

To sum up, what should science and engineering students know about choosing the best databases to search?

1. You can only search the databases that are available to you, and if you are no longer affiliated with a college campus, then your database options might be limited.
2. You'll need to choose the best databases to search based on various dimensions of your information need.
3. For most *scientists*, it's better to search article databases (when available) rather than publisher databases. But for *engineers*, publisher databases may be essential.
4. The number of databases you should search will depend on your topic, how comprehensively you need to search, and the databases available to you. When in doubt, get help from a librarian!

What's Next?

You're done with **Part V** of this book and you've learned all about databases! Next, we'll move on to **Part VI**, where we'll discuss specific strategies for **searching databases**.

PART VI
SEARCHING DATABASES FOR SOURCES

CHAPTER 24
IDENTIFYING KEYWORDS TO SEARCH

Identifying the best keywords to search is one of the most useful (and most underestimated) skills you can learn. This chapter will show you *why* it's so important to select the fewest, most essential keywords that represent your topic—plus you'll learn *how* to select the best keywords. Once you get in the habit of searching with carefully chosen keywords, you'll be able to find information much more efficiently and effectively.

Identifying Strategic Keywords Is the First Step in Conducting a Comprehensive Search

When you're ready to look for sources on a topic, the process you use to search databases will depend on your particular goals. Do you simply need to find a few relevant sources? If so, then pick one or more appropriate databases to search (see Chapters 20 and 23 for help choosing), enter a keyword search, then scan over the first couple of pages of your results. Pay attention to source titles and whether those titles indicate the source contains information relevant to your topic. If you find enough sources that look promising, you probably don't need to conduct more searches. You've found enough without having to plan out a more strategic approach.

But what if you're working on a higher-stakes research project that requires you to search thoroughly for sources? In that case, you'll need to follow a set of steps to ensure your search is robust and comprehensive. Let's take a look at those steps in Table 24.1, including where you can find more information about each step. As the first step in the process is to identify keywords, let's begin by discussing keywords and why they matter.

Table 24.1

Steps in the search process	*Covered in*
1. Identify the fewest, most essential keywords that represent your topic.	This chapter
2. Gather synonyms for your keywords.	Chapter 25
3. Modify your keywords/synonyms with truncation and quotation marks.	Chapter 25
4. Set up a Boolean search.	Chapter 25
5. Choose databases to search.	Chapter 23
6. Run your search, then refine it as needed.	Chapter 26

What is Keyword Searching and Why Is It Important?

When you search databases for information on a topic, you obviously need to enter and search words that represent that topic. We call these **keywords**, but they're also referred to as **search terms**. For simple topics where you're just trying to learn some quick, basic information, it's easy to identify keywords because they're usually the same as your topic. For example, if you need information about the collapse of the Tacoma Narrows bridge or about marine microplastics, you could just enter those keywords into Google or some other database, like this:

- **Tacoma Narrows bridge collapse**
- **marine microplastics**

When your topic is more complex, however, identifying keywords is harder. For instance, suppose this is your research topic:

- **Are there species of freshwater fish that can serve as umbrella species for making conservation decisions?**

(FYI, an umbrella species is a species which, if conserved and protected, will result in the "trickle-down" protection of other species in the same ecosystem.) You might be thinking, "What's so hard about searching that topic? Just enter that sentence into a search box or pull some important words and/or phrases from that sentence and start searching them." It's true this strategy will suffice in certain situations. For example, suppose you're working on a college assignment where you need to find five journal articles on that topic and then write a paper where you cite those journal articles. So, you copy and paste your topic sentence into Google Scholar to conduct a search.

Google Scholar will look at your sentence and try to figure out what the most important keywords are and then return results with those keywords, or with keywords it considers to be synonyms. Those words will be in bold in both the titles and descriptions of your results, as you see in Figure 24.1.

These bold words tell you how Google Scholar is processing your search. You can see it's looking for sources that have words in them like **umbrella**, **species**, and **conservation**—and that's not surprising because those words were in the topic sentence you entered in the Google Scholar search box. But also notice Google Scholar retrieved results with the phrase **keystone species**. That's because it's assuming **keystone species** is a synonym for **umbrella species**, and that you'll want to see results with either phrase. If you look more closely at these results, you'll see that Google Scholar also thought these words and phrases were important, because they are in bold:

- **could**
- **making**
- **available**
- **should**
- **to make**
- **doing**
- **be used**

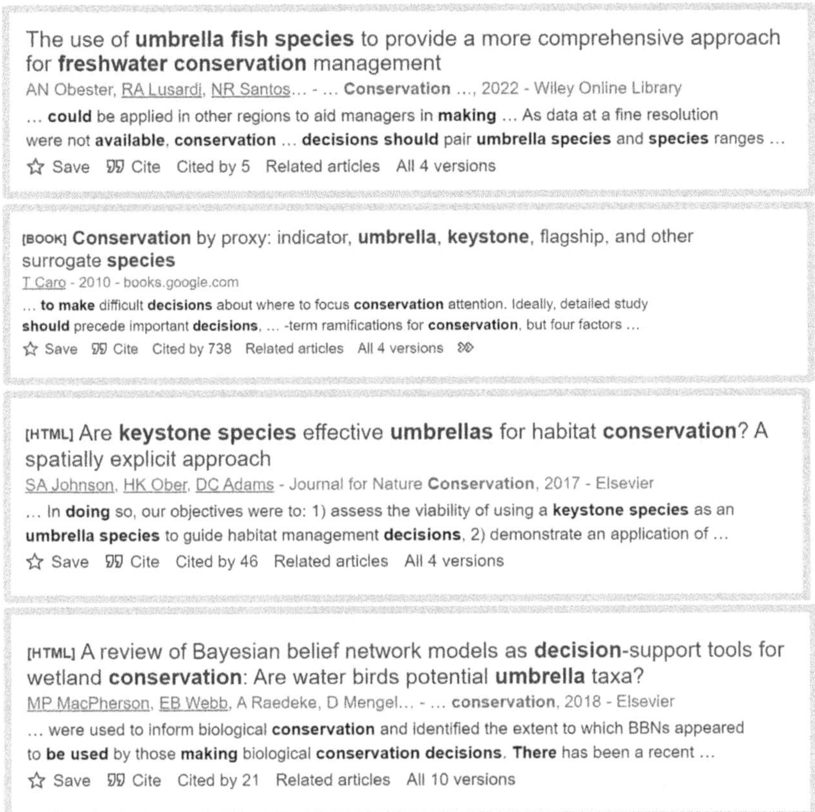

Figure 24.1

Presumably, Google Scholar is retrieving results with those words because it's trying to find synonyms for certain phrases in the sentence you entered, such as the phrases "Are there" and "that can." But you don't want Google Scholar to be searching sources for words and phrases like this. You want it to stay on target and just look for words like **fish**, **umbrella**, and **species**.

When you read the titles of these results, you can see that the first one looks great, but the others don't look so good because they don't appear to focus on fish as umbrella species. That's happening because Google Scholar is flailing around trying to figure out what words in your topic sentence are most important to find. So, these are not optimal results.

The thing is, when you are just starting out doing research, it's hard to know whether the results that your search retrieved are excellent or mediocre, and whether you should keep looking or just work with what you've already found. Also, even if you can tell that your search results aren't great, you may not know it's because your search strategy wasn't great. Instead, you might assume that the information you need just isn't out there.

But for this search on freshwater fish serving as umbrella species, the results are bad because the search strategy is bad. A more strategic keyword search would retrieve vastly better results. However, the reality is that for most college research assignments, you won't be graded on whether you make an effort to search more strategically. In fact, you can usually get by with running a careless search, finding a few semi-okay sources, pulling some content from those sources, and then making that content sound relevant when you write your paper. After all, your instructors will probably grade you on the grammar and coherence of your writing, how well you incorporate sources, whether the sources you used appear to be credible, and if you cited your sources properly. It would be difficult and time-consuming for an instructor to assess how skillfully and thoroughly you searched.

So, typical college research assignments are not ideal for teaching or assessing your ability to conduct strategic, thorough searches for information. College just doesn't give you the practice and feedback you need to improve these skills. In fact, because you can still make good grades on research assignments even when you *don't* conduct strategic, thorough searches, many students graduate thinking they've mastered searching and are expert searchers! That's a perfectly logical conclusion given the circumstances, but the reality is that in college your search skills are never put to a true test.

When you start your career, however, you might work on research projects where you'll need to be able to conduct strategic, thorough searches, and that will require a much more tactical search strategy. So, let's look at a post-college scenario where it would be critical to select the best keywords to search.

Suppose you're employed as a conservation manager for a recreational fishing area (such as a lake or a stretch of river) that's heavily used by both the local community and tourists. Your boss wants you to put together a well-researched conservation management plan for the area and explore the utility of using freshwater fish as umbrella species for making conservation decisions. Furthermore, you'll be posting your conservation management plan online and hosting a community meeting about it, so your plan will be subject to feedback and comments (and probably criticism!) from the local community.

You certainly don't want to go into a public meeting, explain your plan, and then have someone tell you that your plan is flawed and incomplete because it fails to incorporate important research findings from published sources *that you missed in your search for information*! This scenario would be a true test of your search skills because in this example, the stakes are high to conduct a strategic, thorough search so you can put forward a plan that's supported by the best, most relevant research.

An experienced searcher would start researching this topic by identifying the fewest, most essential keywords that represent the topic. Why? Because this search strategy will simultaneously optimize your chances of gathering *relevant* results, but also *comprehensive* results, so that you don't miss anything important. It's the best way to search both thoroughly and effectively.

Knowing this, you decide that the word **fish** as well as the phrase **umbrella species** represent the fewest, most essential keywords. By your reckoning, any sources you retrieve would *have* to have those words, at an absolute minimum, in order to be potentially relevant. It also occurs to you that if you get lots of search results about marine species—rather than freshwater species—then you can always add the keyword **freshwater** to your search, but for now you're going to start as simple as possible to see what you can find.

When you run a search in Google Scholar with just those three keywords, Figure 24.2 shows some of the articles you retrieve.

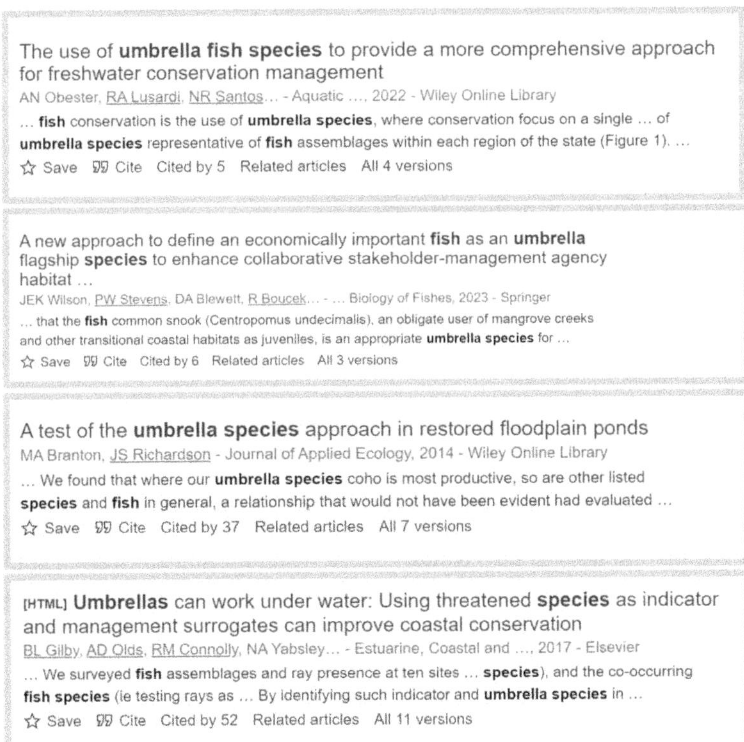
The use of **umbrella fish species** to provide a more comprehensive approach for freshwater conservation management
AN Obester, RA Lusardi, NR Santos... - Aquatic ..., 2022 - Wiley Online Library
... **fish** conservation is the use of **umbrella species**, where conservation focus on a single ... of **umbrella species** representative of **fish** assemblages within each region of the state (Figure 1). ...
☆ Save 99 Cite Cited by 5 Related articles All 4 versions

A new approach to define an economically important **fish** as an **umbrella** flagship **species** to enhance collaborative stakeholder-management agency habitat ...
JEK Wilson, PW Stevens, DA Blewett, R Boucek... - ... Biology of Fishes, 2023 - Springer
... that the **fish** common snook (Centropomus undecimalis), an obligate user of mangrove creeks and other transitional coastal habitats as juveniles, is an appropriate **umbrella species** for ...
☆ Save 99 Cite Cited by 6 Related articles All 3 versions

A test of the **umbrella species** approach in restored floodplain ponds
MA Branton, JS Richardson - Journal of Applied Ecology, 2014 - Wiley Online Library
... We found that where our **umbrella species** coho is most productive, so are other listed **species** and **fish** in general, a relationship that would not have been evident had evaluated ...
☆ Save 99 Cite Cited by 37 Related articles All 7 versions

[HTML] **Umbrellas** can work under water: Using threatened **species** as indicator and management surrogates can improve coastal conservation
BL Gilby, AD Olds, RM Connolly, NA Yabsley... - Estuarine, Coastal and ..., 2017 - Elsevier
... We surveyed **fish** assemblages and ray presence at ten sites ... **species**), and the co-occurring **fish species** (ie testing rays as ... By identifying such indicator and **umbrella species** in ...
☆ Save 99 Cite Cited by 52 Related articles All 11 versions

Figure 24.2

You still get that first article you got before, which is great, but compare the other three results to those you retrieved when you entered your topic into Google Scholar as a full sentence. You can see the results in Figure 24.2 are more targeted and relevant because Google Scholar knows to just look for the words **fish umbrella species**.

To sum up, when you enter a full sentence or long phrases in a search, you're making Google Scholar guess at which words are most important, so it'll flail around a bit in its effort to retrieve relevant results. However, when you pare down your sentence and just enter the most important keywords, Google Scholar can focus on finding results with just those important keywords.

Why Searching the Fewest, Most Essential Keywords Is So Important

You just learned that searching the fewest, most essential keywords works better in Google Scholar because you're being precise about what you want it to find, instead of asking it to try to figure out the best keywords from a long string of words you've entered. Basically, if you don't want Google Scholar to look for results that have unimportant words like **there**, **any**, **would**, etc., then don't enter a search that has those kinds of words. As you saw in the section above, you'll still *get* results when you search this way, but they'll be muddier and less relevant.

For other types of databases, such as the databases offered on campus library websites, searching the fewest, most essential keywords isn't just better, it's *critical*. To see why, take another look at Figure 22.4 to see how the database MoonCites works by retrieving just those records that have every single one of the keywords you asked it to find. Now, imagine entering a long sentence into MoonCites that represents your topic. MoonCites will try to find records that contain every one of the words in your sentence, which means it may not find any matching records at all.

Searching Web of Science works the same. If you enter that search about freshwater fish as an umbrella species *as a sentence* into Web of Science, you get a grand total of zero results. That's not surprising because you told Web of Science it must find records that have every single one of the keywords you entered, and there were no records that contained all of them.

But what if you remove *some* of the words from your topic sentence and just enter the following:

- **freshwater fish serve as umbrella species for conservation decisions**

Again you get a grand total of zero results! There were still no records that contained all those words, so the database couldn't deliver any results. But what happens when you pare down to the bare minimum of words that would need to be present in a source for that source to be potentially relevant to this topic? This time, we'll just enter:

- **fish umbrella species**

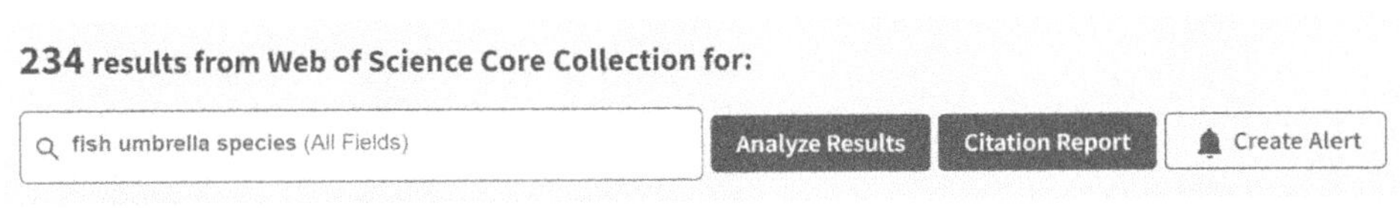

Figure 24.3

Now you get 234 results from Web of Science (see Figure 24.3). Not all of them will be relevant, but *some* will be, and finding as many relevant results as possible would be critical for that employment scenario where your boss asked you to come up with a conservation management plan that's supported by the best research.

To get a visual image of how bossy and demanding it is to ask a database to search an entire sentence or a long string of keywords—instead of searching just a few, essential keywords—compare what you are telling the database to do:

"Retrieve sources that contain the word **are** and the word **there** and the word **species** and the word **of** and the word **freshwater** and the word **fish** and the word **that** and the word **can** and the word **serve** and the word **as** and the word **umbrella** and the word **species** and the word **for** and the word **making** and the word **conservation** and the word **decisions.**"	"Retrieve sources that contain the word **fish** and the word **umbrella** and the word **species.**"

In the second column above, what you're asking the database to find is so much less restrictive! Less-restrictive searches give databases the freedom to find as many relevant results as possible. And by the way, if you're worried that conducting less-restrictive searches will retrieve lots of results to sort through—relax, we will talk about how to manage large result sets in Chapter 26.

To really drive this point home, suppose you were pleased with yourself for paring down your topic sentence to just these four keywords:

- **fish umbrella species conservation**

In Figure 24.4, let's look at how having even *one* non-essential keyword (like the word **conservation**) can cause you to miss out on relevant articles. For two of the articles, the authors happen to be using words like **biodiversity** and **ecological management** *instead* of **conservation** in the titles of their articles. It's true that the word **conservation** might show up somewhere in the full records for these articles (perhaps in the abstract), but maybe not. If not, then the database is not going to retrieve these articles for you. However, if you'd stuck with searching just the fewest, most essential keywords (**fish umbrella species**), you'd have retrieved these highly relevant articles.

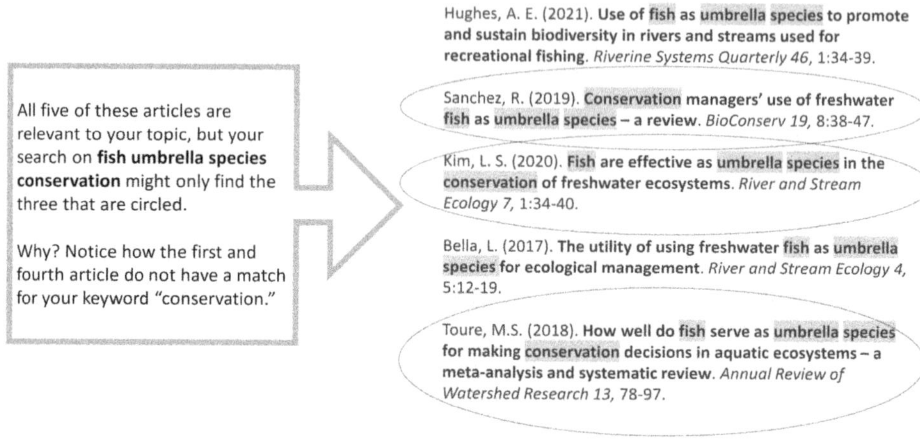

Figure 24.4

To sum up, most databases (other than Google and Google Scholar) are very literal, and they *only* retrieve sources that have *all* the keywords you enter into the search box. So, don't search any more keywords than you absolutely need, or you're likely to miss out on relevant sources.

How to Identify the Fewest Most Essential Keywords for Your Topic

Now let's talk about how to identify the fewest, most essential keywords that represent your topic. For the topic we've been working with, we got good results using just the words **fish umbrella species**, but how did we land on *those* as the best keywords to search? First look at your topic again:

- **Are there species of freshwater fish that can serve as umbrella species for making conservation decisions?**

Then, think about the most *essential* concepts that represent that topic. At the heart of it, you need to find sources about *a type of animal* playing *a type of role*—and that's it. The simplest way to refer to the animal is **fish**. The simplest way to refer to the role is **umbrella species**. In other words, at the very least, a source would need to mention **fish** and it would need to mention **umbrella species** in order to be potentially relevant. So, start your research by searching *only* those keywords.

You can always adapt your search if you get lots of results that are not relevant. For example, if you notice you're getting too many results about marine species instead of freshwater species, then you could consider adding the word **freshwater** to your search to tighten it up, or you could tell the database to exclude sources with the words **marine** or **ocean** (you'll see how to exclude words from your search in the next chapter). But as a general rule, you should start by searching only the fewest, most essential keywords.

Let's practice with another topic:

- **What is the best equipment and design for deploying drones to monitor volcanic gas plumes?**

Jump right in and think about the essence of this topic, which is about using a certain tool (drones) to study a particular phenomenon (volcanic gas plumes). After all, any source that contains those keywords will probably relate to equipment, design, or deployment anyway. So, that's why it's unnecessary to add keywords like "equipment," "design," "deployment," etc., to your search. Instead, pare down and just search:

- **drones volcanic gas plumes**

Keep in mind that some of your keywords might be phrases, and if they are, you can put them in quotation marks to force the database to find that exact phrase. So, if you're searching for sources that discuss aeroelastic issues surrounding the Tacoma Narrows bridge collapse, then you could search Tacoma Narrows as a phrase as you see in this keyword search:

- **"Tacoma Narrows" bridge aeroelastic**

You'll learn more about phrase searching—including its advantages and disadvantages—in the next chapter.

If you'd like to get a little more practice choosing the fewest, most essential keywords that represent a topic, try to pick out the best keywords for these three topics:

- **What is the effect that basalt has on the morphology of streams associated with the Mogollon Rim?**
- **What types of therapies are used to treat arthritis in the hand, and which are the most promising?**
- **How are mobile LiDAR scanners being used to observe changes in forest structure?**

See how well your keyword choices match up with those that are underlined below.

- **What is the effect that <u>basalt</u> has on the <u>morphology</u> of <u>streams</u> associated with the <u>Mogollon Rim</u>?**
- **What types of <u>therapies</u> are used to treat <u>arthritis</u> in the <u>hand</u>, and which are the most promising?**
- **How are mobile <u>LiDAR</u> scanners being used to observe changes in <u>forest structure</u>?**

Is there always one right answer for which keywords best represent your topic? No, there can be ambiguity and you may need to experiment with different searches to see what works best. Take, for example, this topic:

- **How effective are Arizona's highway underpasses at helping elk cross highways safely?**

Both of the pared-down keyword searches below will retrieve good results:

- **Arizona underpasses elk**
- **Arizona highways elk**

So, just note that sometimes there's more than one good set of essential keywords.

What If You Don't Have a Nicely Worded Topic like Those We've Been Discussing?

Then work on stating your topic as a sentence or question! As mentioned in Chapter 19, creating a topic statement helps you stay focused on the precise information you need to find when you start searching for sources, plus it helps you identify the fewest, most essential keywords that represent your topic. Refer to Chapter 19 for tips on refining your topic and stating it in a single sentence.

To sum up, what do science and engineering students need to know about identifying keywords?

1. Identifying the fewest, most essential keywords that represent your topic is the most strategic way to start your research. It will simultaneously optimize your chances of gathering *relevant* results, but also *comprehensive* results.
2. To identify the fewest, most essential keywords for your topic, consider which concepts in your topic capture its essence.

What's Next?

Identifying the fewest, most essential keywords is really just the first step in conducting a good search. Next, you'll need to think about adding **synonyms**, modifying your keywords/synonyms with **truncation** and **quotation marks**, and then constructing a **Boolean search**. We'll tackle those skills in the next chapter.

CHAPTER 25
CONDUCTING COMPREHENSIVE SEARCHES

The skills explained in this chapter are particularly useful for conducting *comprehensive* searches. You'll learn how to expand your searches with synonyms and truncation—plus you'll learn how to leverage phrase searching and Boolean searching. These skills will help you become a more efficient and effective searcher, even at times when you don't need to conduct a particularly thorough search.

What's So Special about the Skills Discussed in This Chapter?

While the skills explained in this chapter will help you search more comprehensively, it's not always necessary to do comprehensive searches. It really depends on the circumstances. For example, most college research assignments don't require you to conduct comprehensive searches—instead, you're usually asked to find just five to ten relevant sources. But this never happens in real life. Recall that employment scenario from the last chapter where your boss wanted you to put together a well-researched conservation management plan for a recreational fishing area? Your boss would never say: "Just find five sources on this topic and develop your conservation management plan from those five sources—and simply disregard any other relevant sources." However, this is exactly what you are taught to do for college assignments, which is why those assignments don't reflect reality! So, if you want to learn skills for comprehensive searching that you can put to use in your career, then this chapter is for you. Even if you don't think you'll *ever* need to conduct a comprehensive search, just being aware of the skills in this chapter will help you find information with less time and effort.

Gathering Synonyms Will Make Your Search More Comprehensive

In this chapter, we'll focus on the second, third, and fourth steps of the search process (see Table 25.1), starting with gathering synonyms for your keywords. Recall in the last chapter you learned to start your search with the fewest, most essential keywords in order to optimize your chances of finding as many relevant sources as possible. But to find *even more* relevant sources, you'll want to add **synonyms** to your search. To better understand this, let's take another look at an example topic from the last chapter:

- **What is the best equipment and design for deploying drones to monitor volcanic gas plumes?**

Table 25.1

Steps in the search process	*Covered in*
1. Identify the fewest, most essential keywords that represent your topic.	Chapter 24
2. Gather synonyms for your keywords.	This chapter
3. Modify your keywords/synonyms with truncation and quotation marks.	This chapter
4. Set up a Boolean search.	This chapter
5. Choose databases to search.	Chapter 23
6. Run your search, then refine it as needed.	Chapter 26

If you recall, we decided on these as the fewest, most essential keywords:

- **drones volcanic gas plumes**

Now consider whether a search using those keywords would retrieve this highly relevant source:

> Rialto, D. (2012). Design of a UAV for volcanic gas plume sampling. *2012 International Conference on Unmanned Technologies* (pp. 12–16). ICUT Publishers.

Your keywords probably won't retrieve that source because the author isn't using the word "drones," they're using a synonym for drones: **UAV**. "UAV" stands for "unmanned aerial vehicle" and it's not the only synonym for drones, there are a few others; some are phrases and some are initials for phrases. Here's a list:

- unmanned aerial vehicles
- unmanned aerial systems
- unoccupied aircraft systems
- remotely piloted aerial systems
- UAV
- UAS
- RPAS

As you now know, most databases are very literal and will *only* retrieve sources that have the *exact* keywords you entered into your search. So, if you ask a database to find sources with the word **drones**, then your search *won't* retrieve sources where an author happens to be using a synonym for drones, such as **UAV**. (Google Scholar is an exception—if you enter the word **drones** into a search it may be able to find some synonyms.)

Now that you know there are lots of synonyms for drones, you'll need to conduct more searches to search this topic thoroughly. In fact, to ensure you find all the relevant sources out there, you'll need to conduct eight searches using all the different ways that people refer to drones:

1. **drones** volcanic gas plumes
2. **unmanned aerial vehicles** volcanic gas plumes

3. **unmanned aerial systems** volcanic gas plumes
4. **unoccupied aircraft systems** volcanic gas plumes
5. **remotely piloted aerial systems** volcanic gas plumes
6. **UAV** volcanic gas plumes
7. **UAS** volcanic gas plumes
8. **RPAS** volcanic gas plumes

If you come across *additional* synonyms for **drones**, then you should conduct searches using those synonyms as well. (And there are, in fact, many more synonyms for drones, but we'll just experiment with the eight listed above.) Don't worry, later in this chapter we'll talk about how to find relevant synonyms for your keywords. But for now, let's move on to another, related way to expand your search.

Adding Word Variations to Further Expand Your Search

You've learned that authors might use a variety of synonyms to refer to drones, but also consider that authors might use different variations of the word "volcanic." Specifically, they might use word variations like **volcano** or **volcanoes**, like in this article:

> Contreras, P. (2021). Construction of a heat resistant drone to monitor gas plume emissions from an active volcano. *Journal of Instrument Engineering*, 54(3): 1–21.

Word variations are kind of like synonyms; because you don't know exactly which word variation an author might have used, you'll need to search *all of them* in order to be thorough. But when you do, the number of searches you need to conduct starts to get unwieldy, as you see in Table 25.2. This table shows twenty-four different searches that

Table 25.2

Searches using volcanic	*Searches using* volcano	*Searches using* volcanoes
1. drones volcanic gas plumes	9. drones volcano gas plumes	17. drones volcanoes gas plumes
2. unmanned aerial vehicles volcanic gas plumes	10. unmanned aerial vehicles volcano gas plumes	18. unmanned aerial vehicles volcanoes gas plumes
3. unmanned aerial systems volcanic gas plumes	11. unmanned aerial systems volcano gas plumes	19. unmanned aerial systems volcanoes gas plumes
4. unoccupied aircraft systems volcanic gas plumes	12. unoccupied aircraft systems volcano gas plumes	20. unoccupied aircraft systems volcanoes gas plumes
5. remotely piloted aerial systems volcanic gas plumes	13. remotely piloted aerial systems volcano gas plumes	21. remotely piloted aerial systems volcanoes
6. UAV volcanic gas plumes	14. UAV volcano gas plumes	22. UAV volcanoes gas plumes
7. UAS volcanic gas plumes	15. UAS volcano gas plumes	23. UAS volcanoes gas plumes
8. RPAS volcanic gas plumes	16. RPAS volcano gas plumes	24. RPAS volcanoes gas plumes

could all return relevant results. If you ran all these searches, you could be fairly certain you'd conducted a comprehensive search on this topic.

You might be wondering, "When I search words like **volcanic**, doesn't the database know to look for versions of that word with different endings?" Many databases can find very simple singular/plural/possessive forms of words where the presence or absence of an "s" is the only difference in word forms. So, you can be confident that a search on **drone** will capture results with the word **drones**. Likewise, a search on **drones** will capture results with the word **drone**. But most databases can't find variable word endings much more complex than that.

You might also be panicking and wondering "Do I really need to conduct twenty-four different searches in order to search this topic comprehensively?" Thankfully, no! There are a couple of cool techniques you can employ to drastically reduce the number of searches you need to conduct, yet still feel confident you've run a comprehensive search. One of those techniques is called "truncation," so let's start with that.

Truncation

Truncation means removing part of something to shorten it. Truncation is also sometimes called **stemming**. You can use this technique whenever you want a database to find a word with all its different endings. For example, you can truncate the word **volcanic** so the database will retrieve any variation of this word including **volcano**, **volcanoes**, and **volcanism**. All you do is shorten the word at the last letter that all the variations have in common, and then add an asterisk at the end, like this: **volcan***. Using this technique, you can eliminate the need to run all twenty-four of the searches in the table above, and just dial it back to eight again:

1. drones **volcan*** gas plumes
2. unmanned aerial vehicles **volcan*** gas plumes
3. unmanned aerial systems **volcan*** gas plumes
4. unoccupied aircraft systems **volcan*** gas plumes
5. remotely piloted aerial systems **volcan*** gas plumes
6. UAV **volcan*** gas plumes
7. UAS **volcan*** gas plumes
8. RPAS **volcan*** gas plumes

But, there are two important things to be aware of when truncating keywords. First, be careful *where* you add the asterisk to capture different word endings. Suppose you were conducting a search about techniques for composting industrial waste. One of your keywords will be **composting** but you'll also want to capture word variations like **compost**, **composted**, and **compostable.** So, you enter **comp*** into your search to gather all these variable word endings. Unfortunately, that search will retrieve sources that have unrelated words with the same prefix, such as: **computer**, **composition**, **comparison**, **compromise**, **compadre**, **company**, **companion**, etc. So, be careful about shortening your keywords too much; in this example, the search **compost*** would be more strategic.

Second, note that truncating words with an asterisk works in nearly all databases except Google and Google Scholar. Those databases can't process words like **volcan*** or **comput*** or anything else with an asterisk. Instead, you have to enter full words.

There's another technique (besides truncation) that you can use to reduce the number of searches you need to conduct. It's called Boolean searching, so let's tackle that next.

Boolean Searching

The phrase "Boolean searching" sounds technical, but it's not. **Boolean searching** is a type of search strategy that's named after the man who developed and popularized it, George Boole. So, it's called "Boolean searching" because it's George Boole's style of searching—just as if this style of searching had been developed and popularized in the state of California, it might have been called "Californian searching."

Boolean searching offers a way to take these eight searches and reduce them to just one:

1. drones volcan* gas plumes
2. unmanned aerial vehicles volcan* gas plumes
3. unmanned aerial systems volcan* gas plumes
4. unoccupied aircraft systems volcan* gas plumes
5. remotely piloted aerial systems volcan* gas plumes
6. UAV volcan* gas plumes
7. UAS volcan* gas plumes
8. RPAS volcan* gas plumes

To learn how to do that, you'll need to start entering keywords into databases a little bit differently than you're used to. You're probably accustomed to entering your keywords all in one box, like the search you see in the database Web of Science (Figure 25.1). After

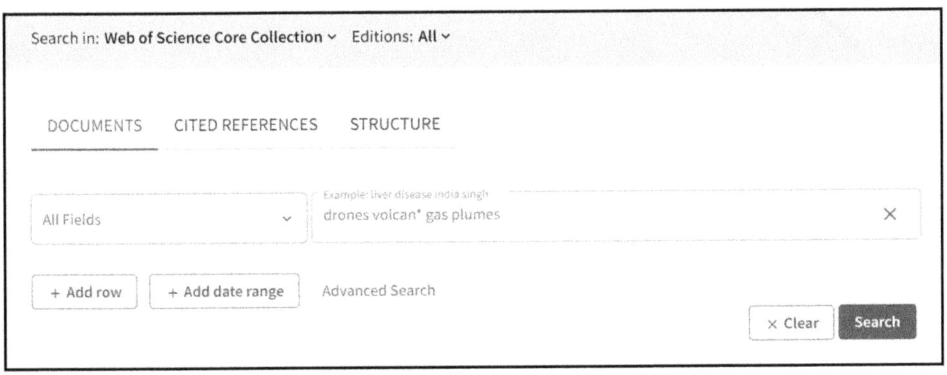

Figure 25.1

all, this is the way you're accustomed to searching Google and Google Scholar as well—by putting all your keywords in one box.

When you enter searches this way, the database processes your search as if you had entered the word **AND** between each keyword, as shown in Figure 25.2.

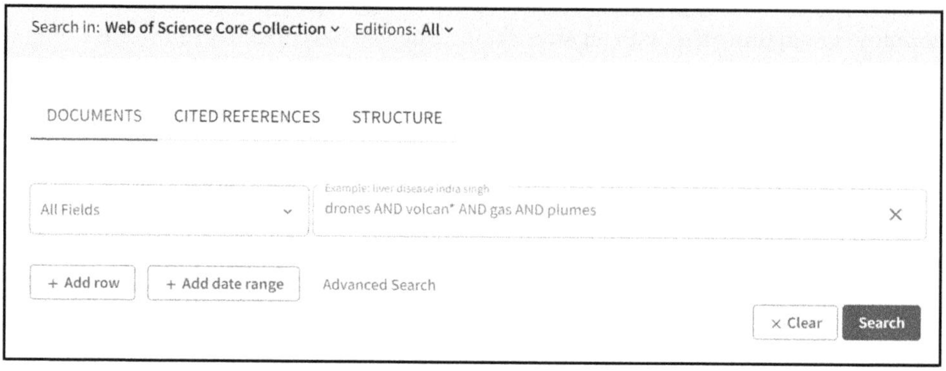

Figure 25.2

In other words, the database interprets both these searches exactly the same way:

- **drones volcan* gas plumes**
- **drones** AND **volcan*** AND **gas** AND **plumes**

Here's what both of those searches are asking the database to do in the form of a command: "Find sources that contain the word **drones** and any word that starts with '**volcan**' and the word **gas** and the word **plumes**."

But, there are other ways to enter that same search into a database. For example, you can use the **+Add row** button (Figure 25.3) to add more search boxes that are connected

Figure 25.3

by "And." Or, you could enter the search with **drones** in its own box and **volcan* gas plumes** in its own box and it would achieve the same thing (Figure 25.4).

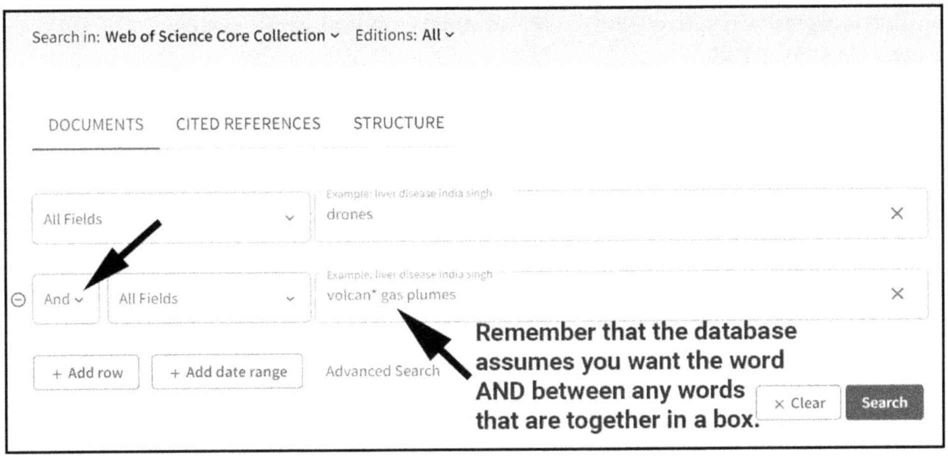

Figure 25.4

So, all four of these different ways of entering the search are functionally the same and would retrieve the same results. To sum up, the default for nearly all databases is to assume you want your keywords connected by the word **AND**, even if you don't actually type the word **AND** between your keywords.

If all these searches achieve the same thing, then why bother to put some or all of your keywords in their own individual box, rather than just entering them all into the same box? Because when you separate a keyword into its *own* box, then you can add synonyms for that keyword. For example, by putting the word **drones** in its own box, we can now tell the database to find the word "drones" or any of its synonyms (see Figure 25.5).

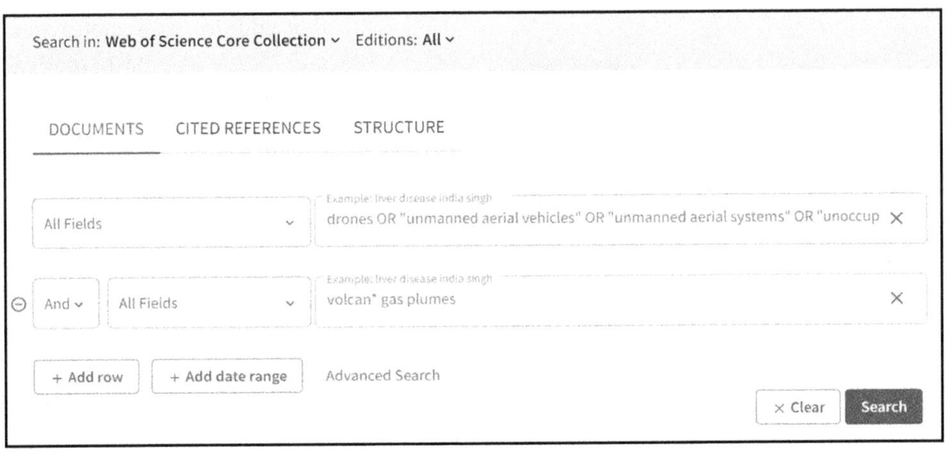

Figure 25.5

So, in that top box we're telling the database to find sources with at least one of these words or phrases, but we don't care which one:

> **drones** OR **"unmanned aerial vehicles"** OR **"unmanned aerial systems"** OR **"unoccupied aircraft systems"** OR **"remotely piloted aerial systems"** OR **UAV** OR **UAS** OR **RPAS**

When you enter the search this way, you reduce eight searches to just one search (see Table 25.3). That search in the second column is a Boolean search and it's called that because it uses **Boolean operators**, which you'll learn about in a moment. But first, you may have noticed when we condensed those eight separate searches into just one Boolean search, we put phrases like **unmanned aerial vehicles** inside quotation marks, just as we did for the phrase "Tacoma Narrows" in a search you saw in the last chapter. We've not gotten into much depth yet about phrase searching, but we will shortly.

Table 25.3

Eight searches	*One search*
1. drones volcan* gas plumes 2. unmanned aerial vehicles volcan* gas plumes 3. unmanned aerial systems volcan* gas plumes 4. unoccupied aircraft systems volcan* gas plumes 5. remotely piloted aerial systems volcan* gas plumes 6. UAV volcan* gas plumes 7. UAS volcan* gas plumes 8. RPAS volcan* gas plumes	1. drones OR "unmanned aerial vehicles" OR "unmanned aerial systems" OR "unoccupied aircraft systems" OR "remotely piloted aerial systems" OR UAV OR UAS OR RPAS AND volcan* gas plumes

Boolean Operators

Perhaps you've heard people refer to "Boolean operators" and you've wondered what that's all about. Remember when you were learning math in grade school, and you were taught that addition and subtraction are called **mathematical operations** and the plus and minus symbols are referred to as **operators**? Well, when you search using **AND**, **OR**, and **NOT**, those words are considered to be operators as well, hence the name **Boolean operators**. You've already seen Boolean searches that incorporate **AND** and **OR**, and in the next section you'll see how to use **NOT**.

You might also be wondering why you see Boolean operators capitalized in Boolean searches. When you capitalize Boolean operators, that tells the database to process the word "or" as an operation (that is, as a command to do something) rather than to search for the word "or" and retrieve sources that have that word. Actually, nowadays many databases know to interpret a lower case "or" as a Boolean operator because—after all—who would ever want to search the word "or"? But not all databases are this progressive, so it's better to capitalize your Boolean operators to ensure the database interprets them correctly.

More Practice with Boolean Searching

Let's get a little more practice with Boolean searching by experimenting with a topic we explored in the previous chapter:

- **Are there species of freshwater fish that can serve as umbrella species for making conservation decisions?**

If you recall, in the last chapter we settled on **fish umbrella species** as the fewest, most essential keywords to search. Let's suppose we wanted to expand this search to look for sources where fish were studied in a variety of conservation roles, including **umbrella**, **flagship**, **indicator**, or **keystone** species. *And*, we want to eliminate results that are about **marine** or **oceanic** fishes. You can eliminate specific words from your search by utilizing the Boolean operator **NOT**. In Figure 25.6 you see how to enter the search with **marine** and **ocean*** eliminated. The word "ocean" is truncated to capture both **ocean** or **oceanic**.

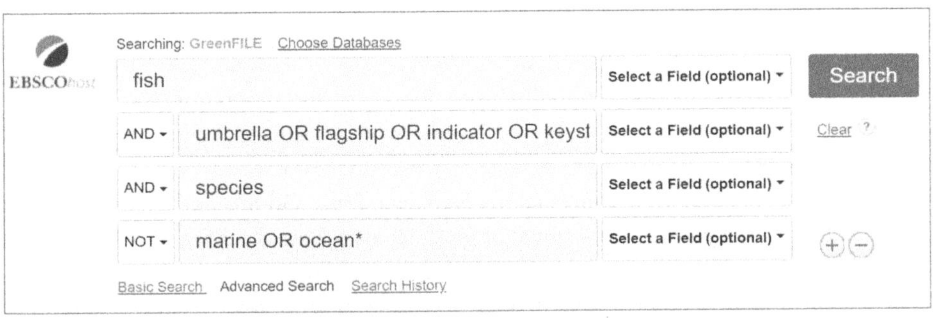

Figure 25.6

If you decide to eliminate certain words from your search using **NOT**, then note that those words always need to be in the *bottom search box*. As for your other keywords, the order you enter them into the stacked search boxes doesn't matter. In other words, all of the searches shown in Table 25.4 would retrieve the same results.

Table 25.4

fish AND **umbrella** OR **flagship** OR **indicator** OR **keystone** AND **species** NOT **marine OR ocean***	**umbrella** OR **flagship** OR **indicator** OR **keystone** AND **species** AND **fish** NOT **marine OR ocean***	**species** AND **fish** AND **umbrella** OR **flagship** OR **indicator** OR **keystone** NOT **marine OR ocean***

How to do Boolean Searching in Google and Google Scholar

As you've learned in this chapter, putting a keyword with its synonyms in its own box, with each synonym separated by **OR**, allows you to conduct a sophisticated, highly effective, Boolean-style search. But Google and Google Scholar don't offer multiple, stacked search boxes like other databases do. If you wanted to run a search on how forest fires negatively affect water quality in California streams – and you wanted to include relevant synonyms – then you could enter it as you see in Figure 25.7.

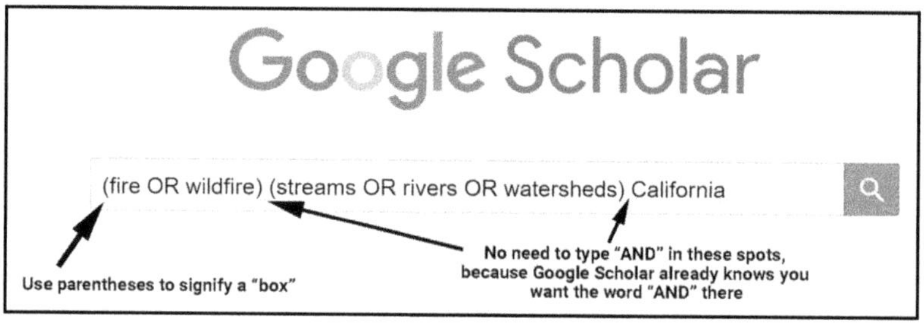

Figure 25.7

Note that any word on its own—for which you don't have any synonyms to add—doesn't need to be in parentheses. That's why you see the word "California" on its own and not inside parentheses.

Just be aware that when it comes to Boolean searching, Google and Google Scholar can't handle as much complexity as most other databases. Keep your searches simple and don't enter dozens of synonyms or lots of different groupings separated by parentheses. If you do, these databases will fail to process your search properly.

Phrase Searching 101

When you're conducting a Boolean search, and some of your keywords are phrases, then you must put them inside quotation marks. **Quotation marks** tell the database to find those words together as a unit. To understand why this is important, suppose you entered your keywords without using quotation marks, like this:

drones OR **unmanned aerial vehicles** OR **unmanned aerial systems** OR **unoccupied aircraft systems** OR **remotely piloted aerial systems** OR **UAV** OR **UAS** OR **RPAS**

Now remember that when a database sees strings of keywords with nothing between them, it defaults to adding the word **AND** between them, like this:

drones OR **unmanned** AND **aerial** AND **vehicles** OR **unmanned** AND **aerial** AND **systems** OR **unoccupied** AND **aircraft** AND **systems** OR **remotely** AND **piloted** AND **aerial** AND **systems** OR **UAV** OR **UAS** OR **RPAS**

The database is not going to know how to process a search with all those ANDs and ORs mixed together like that! However, when you tell the database to process phrases as their own units by putting them in quotation marks, it will be able to process the search properly:

drones OR **"unmanned aerial vehicles"** OR **"unmanned aerial systems"** OR **"unoccupied aircraft systems"** OR **"remotely piloted aerial systems"** OR **UAV** OR **UAS** OR **RPAS**

Keep in mind, though, when you ask a database to find words within quotation marks, you're asking it to find *exactly* those words, and those words *only*, in *exactly* that order. That means a search on the plural phrase **"unmanned aerial vehicles"** may not retrieve instances where an author is using the singular phrase **"unmanned aerial vehicle."** So, that search we constructed earlier in this chapter was not quite right:

drones OR **"unmanned aerial vehicles"** OR **"unmanned aerial systems"** OR **"unoccupied aircraft systems"** OR **"remotely piloted systems"** OR **UAV** OR **UAS** OR **RPAS**

AND **volcan* gas plumes**

Can you see what's wrong with it? We should have made sure we were searching phrases in both their singular and plural forms. Below, the search has been corrected with the singular forms of phrases added:

drones OR **"unmanned aerial vehicles"** OR **"unmanned aerial vehicle"** OR **"unmanned aerial systems"** OR **"unmanned aerial system"** OR **"unoccupied aircraft systems"** OR **"unoccupied aircraft system"** OR **"remotely piloted systems"** OR **"remotely piloted system"** OR **UAV** OR **UAS** OR **RPAS**

AND **volcan* gas plumes**

By the way, in the last chapter you learned that most databases can find simple singular and plural forms of a word—and that's still true, *except* if the word (or phrase) is inside quotation marks. In general, quotation marks tell the database to only look for *exactly* what's inside the quotation marks, with no variation allowed. So, when you're searching for phrases, consider whether you need to search that phrase in both its singular and plural forms.

You might be wondering about using a combination of asterisks and quotation marks to capture singular and plural forms of phrases, like this:

> **drones** OR **"unmanned aerial vehicle*"** OR **"unmanned aerial system*"** OR **"unoccupied aircraft system*"** OR **"remotely piloted aerial system*"** OR **UAV** OR **UAS** OR **RPAS**

Be aware that many databases will not process that search correctly. Why not? The database is looking for *exactly* what's inside the quotation marks so it will be looking for phrases in which the last word has an asterisk at the end; the database won't understand you want to find an exact phrase, but one that has different word endings. So, don't mix asterisks with quotation marks unless you know you're using a database that's capable of interpreting that sort of search properly.

Phrase Searching 102

Now let's get more in-depth with phrase searching and talk about some advantages and disadvantages you should be aware of. Suppose you're searching for information about **forever chemicals** in the database Web of Science. When you conduct your search, you get 263 results. As you start to look through these results, you're surprised by record number 176, shown in Figure 25.8.

176 The First Russian Mineralogist - Nikolai Ivanovich Koksharov (1818-1892)
Rudnik, SN; Voloshinova, IV; (...); Leyberov, AI
2019 | BYLYE GODY 54 (4) , pp.1546-1553
This article is devoted to the study of the biography of the outstanding Russian scientist Nikolai Ivanovich Koksharov. He was one of the largest Russian mineralogists. For his contribution to the development of this field of mining engineering, N.I. Koksharov earned the ... Show more
4 Citations
15 References

Figure 25.8

It certainly doesn't look very relevant to the topic **forever chemicals**! So, why was this result retrieved? When you click on it to open the full record, you'll see that both the words **forever** and **chemical** appear in the abstract of the record. Here's a quote from the abstract, with the relevant words bolded, so you can see for yourself:

> This article is devoted to the study of the biography of the outstanding Russian scientist Nikolai Ivanovich Koksharov. He was one of the largest Russian mineralogists. For his contribution to the development of this field of mining engineering, N.I. Koksharov earned the title "father of Russian crystallography." This inscription was even placed on his grave as a tribute to the scientist, whose name is **forever** inscribed in the history of Russian science. Nikolai Ivanovich became the founder of a new scientific field in Russia, replacing the old method of qualitative description of minerals with mathematically accurate crystallographic study, **chemical** analysis, and physical research.

The words "forever" and "chemical" are present in this record, which is why this record was retrieved, but those words don't show up together as a phrase. So now let's run the search again and look for **forever chemicals** as a phrase. We'll search both the singular and plural forms, like this:

- **"forever chemicals"** OR **"forever chemical"**

That search retrieves far fewer results (ninety-eight rather than 263). That's because we eliminated all the results where the words **forever** and **chemical** show up, but not together as a phrase. Now we have fewer results to look through, and it's likely most or all of them will be relevant. So in this example, phrase searching is a really strategic way to eliminate non-relevant results.

But problems can arise if you don't think carefully about how best to utilize quotation marks for phrase searching. For example, a common mistake is to put too many words inside quotation marks. Suppose you're searching for sources about the humpback chub in Grand Canyon National Park. It would *not* be a good idea to search using these keywords:

"humpback chub"

AND **"Grand Canyon National Park"**

That search would fail to retrieve this highly relevant result:

Kaeding, L. R. & Zimmerman, M. A. (1983). Life history and ecology of the humpback chub in the Little Colorado and Colorado rivers of the Grand Canyon. *Transactions of the American Fisheries Society*, *112*(5): 577–594.

Why? Because you told the database to *only* find the really long phrase "Grand Canyon National Park" and you can see from the title of this source that the authors are using the shorter, more commonly used phrase "Grand Canyon." So, your search won't retrieve this source unless the entire phrase "Grand Canyon National Park" shows up somewhere in the full record, which it probably doesn't.

Something else to note about quotation marks is that you should only use them when you are trying to capture standard phrases like **Grand Canyon**, **forever chemicals**, or **unmanned aerial vehicles**—in which people never deviate from using those exact words in that exact order. Let's look at what happens when you put a non-standard phrase in quotation marks, using this topic as an example:

- **What is the carbon capture capacity of arid rangeland soils?**

Suppose these were the keywords you chose to search:

"carbon capture"

AND **rangeland**

AND **soils**

The problem is that "**carbon capture**" is not a standard phrase that everyone uses without any variation. Authors often use **CO2** instead of **carbon**, plus there are lots of ways to refer to the capture of carbon (**sequestration**, **uptake**, etc.). Not only that, the order of the words can vary (**capturing carbon**)—or, other words could be present in the phrase (**capture of carbon**). Here's a list of possible phrases authors might use to talk about carbon capture:

- carbon/CO2 capture
- carbon/CO2 storage
- carbon/CO2 sink
- carbon/CO2 sequestration
- carbon/CO2 uptake
- storage of carbon/CO2
- storing carbon/CO2
- uptake of carbon/CO2
- sequestering carbon/CO2
- the sequestering of carbon/CO2
- capturing carbon/CO2
- the capture of carbon/CO2
- the capturing of carbon/CO2
- how to capture carbon/CO2
- how to sequester carbon/CO2

So "**carbon capture**" is not a standard phrase universally used by everyone without variation. That means you shouldn't search it as a phrase. Instead, a better way to search this topic would be to separate concepts into stacked search boxes and add synonyms like this:

carbon OR **CO2**

AND **stor*** OR **captur*** OR **sink** OR **uptake** OR **sequester***

AND **rangeland**

AND **soils**

How to Gather Synonyms

At the beginning of this chapter, you saw how you can add synonyms to your search to make it as comprehensive as possible. Then, you saw how to incorporate synonyms into a Boolean-style search. But now we're going to backtrack a bit and discuss how to *gather* synonyms, which is what Step 2 in the research process is all about.

To start, put your keywords in a table, like you see below, and determine if any need to be truncated to capture different word endings. If so, truncate those words (unless you're searching Google or Google Scholar, because those databases can't process truncated words). Then, try to think of synonyms for your keywords, and list them. For example, perhaps you already know that **unoccupied aircraft systems** and the initials for this phrase (**UAS**) are synonyms for drones. Table 25.5 shows how your table would look as you start out.

Table 25.5

drones	*volcan**	*gas plumes*
unoccupied aircraft systems	volcanic	
UAS	volcanoes	
	volcano	

Then, start looking for synonyms. There are several ways to do this. A great way to start is to pick some databases and run searches with the keywords and synonyms you've gathered so far. Once you've run your search, scan through your results and look at source titles to see if you come across any more synonyms. For example, if you searched Google Scholar for **drones volcanic gas plumes** then one of the results you'd get is shown in Figure 25.9. In the title of this result, you can see the word **UAV** is being used instead of the word **drone**. Further investigation would reveal that **UAV** stands for **unmanned aerial vehicle**, so you now have two more synonyms for the word drone.

Remote sensing and identification of **volcanic plumes** using fixed-wing UAVs over Volcán de Fuego, Guatemala
B Schellenberg, T Richardson, M Watson… - Journal of Field …, **2019** - **Wiley** Online Library
… **volcanic plume** on multiple occasions across a range of different conditions. Volcán de Fuego is an active **volcano** which emits **gas** … results of multiple **plume** interceptions with multiple …
☆ Save 99 Cite Cited by 27 Related articles All 10 versions

Figure 25.9

As you find new synonyms (like **UAV** and the corresponding full phrase **unmanned aerial vehicle**) add them to your list (see Table 25.6).

Table 25.6

drones	*volcan**	*gas plumes*
unoccupied aircraft systems	volcanic	
UAS	volcanoes	
UAV	volcano	
Unmanned aerial vehicle		

Another way to find synonyms is by mining the subject terms of records you retrieve in a search. For example, suppose you were messing around with that search shown in Figure 25.7 on how forest fires negatively affect water quality in California streams. You decide to search the keywords **fires streams California** in Academic Search Complete, and when do you retrieve the results shown in Figure 25.10. Notice that these results present you with potentially useful synonyms such as **wildfire** and **watersheds**. So, your search would start to take shape as follows:

fires OR **wildfire**

AND **streams** OR **watersheds**

AND **California**

Figure 25.10

When you find new synonyms, add them to your growing list to keep track of them. Then, run more searches incorporating the new synonyms you found, and scan the titles and subject terms of your results to locate even more synonyms. Repeat this process until you aren't seeing any new synonyms. It's not a bad idea to try this process in a few different databases because you tend to find different synonyms in different databases.

You can also look up keywords (such as "drones") in Wikipedia to see if the Wikipedia entry leads you to new synonyms. Or, look up your keywords in an online thesaurus (such as Thesaurus.com) to find synonyms. In Chapter 23 you saw that you could ask **Generative Artificial Intelligence (AI)** services like **ChatGPT** to get ideas for which databases to search on your topic (although often the advice isn't great). You can also ask these services to generate lists of synonyms. The synonyms for **drones** provided by three different Generative AI Services are shown in Table 25.7. Note that Bing Chat completely misunderstood what was meant by "drones" so you could ask it again but use a phrase like unmanned aerial vehicle instead.

Table 25.7

ChatGPT	*Bing Chat*	*Bard*
Unmanned Aerial Vehicle (UAV) Unmanned Aircraft System (UAS) Remotely Piloted Aircraft (RPA) Aerial Robot Autonomous Aircraft Remote-controlled Aircraft Aerial Drone Flying Robot Quadcopter Multirotor Aerial Vehicle Aerial Device Remote Aircraft Unmanned Aircraft Flying Drone Aerial Quadcopter Airbot Miniature Aircraft Remote Drone Unmanned Flying Device Aerial Apparatus Automated Aircraft Quadrotor Aerial Unmanned System (AUS) Uncrewed Aircraft Aerial Remote-controlled Device Flying Apparatus Autonomous Aerial Vehicle (AAV) Aerial Unmanned Vehicle (AUV) Unpiloted Aircraft	**Nouns**: hum, buzz, whir, vibrate, murmur, intone, pontificate, speak boringly, speak monotonously, go on and on. **Verbs**: make a continuous low humming sound, speak tediously in a dull monotonous tone, move with a continuous humming sound.	Unmanned aerial vehicles (UAVs) Remotely piloted vehicles (RPVs) Unmanned aerial systems (UASs) Remotely operated vehicles (ROVs) Autonomous aerial vehicles (AAVs) Flying robots Drones Quadcopters Octocopters Hexacopters VTOLs (vertical takeoff and landing vehicles) UAV swarms

Last but not least, you could skip all this advice about gathering synonyms and simply consult a librarian who can help you put together a comprehensive list. Librarians are experts at coming up with the most appropriate synonyms, and they have experience knowing which words are likely to be fruitful and which are not. Or you can generate a list of synonyms using Generative AI and/or other methods, and then have a librarian check over the list and streamline it.

In summary, your objective is to gather a comprehensive, robust list of synonyms before assembling and conducting your final search. But, what if you can't think of—or find—any synonyms for some or all of your keywords? That's fine! Some words just don't have synonyms.

Synonyms: A Few More Considerations

On occasion, adding certain synonyms can make your search messier and you may find you need to remove them. For example, suppose you were researching this topic:

- **How can computers be leveraged to teach students math skills?**

Obviously the word **computer** will be one of your keywords, so you decide to look for synonyms for it. You might land on **PC** (which stands for **personal computer**) as a synonym and add it to your search (as in: **computer** OR **PC**). However, that particular synonym also represents the initials for the phrase **politically correct**. So, adding **PC** to your search might deliver a bunch of results that are about political correctness. You may need to remove synonyms from your search if they start to cause problems like this.

Something else to consider when collecting synonyms is that they don't have to be direct equivalents to your main keywords. For example, suppose you were a professional forester researching this topic:

- **What reforestation treatments are most effective for restoring ponderosa pine forests?**

(In forestry, a "treatment" is an activity to achieve a particular outcome.) Perhaps you'd start by experimenting with these main keywords:

Reforestation

AND **treatments**

AND **ponderosa**

You might think to yourself "Are there synonyms for **treatments** that I should be adding to this search?" Good question. If you were researching a *medical topic* then there are obvious synonyms for the word **treatments** such as **therapy** OR **therapies** OR **therapeutics**; all those words are somewhat equivalent in meaning and are often used interchangeably. But in the discipline of forestry, there aren't any words directly interchangeable with the word **treatments**. Instead, forestry researchers often write about specific *types* of reforestation treatments such as **site preparation**, **seeding**, and **planting**. If you were researching this topic, presumably you'd be interested in sources that talk about reforestation treatments generally, as well as specific types of treatments. Assuming that's the case, you could set up some stacked search boxes in a database and enter your search like this:

reforestation

AND **treatment** OR **"site preparation"** OR **seeding** OR **planting**

AND **ponderosa**

Even though the words in your second box are not true synonyms of each other because they don't mean equivalent things, you're still interested in sources that have any one of them. So, you can just treat them *as if* they're synonyms.

It's nuanced situations like these that librarians are good at navigating—and that Generative AI tools are terrible at! That's why consulting a librarian can be really helpful when compiling synonyms.

To sum up, what do science and engineering students need to know about conducting comprehensive searches?

1. You can make your search more comprehensive and find *even more* sources by adding synonyms to your search.
2. To find synonyms, conduct database searches with your keywords and look in the titles and subject terms of your results to get ideas, and/or consult Wikipedia, and/or consult an online thesaurus, and/or use Generative AI.
3. Don't forget to use truncation to capture different word endings (except when searching Google or Google Scholar).
4. It may help to put quotation marks around keywords/synonyms that are phrases, but be strategic about it.
5. You'll save time and effort, and conduct a more comprehensive search, if you organize your keywords and their synonyms into a Boolean-style search.
6. Putting together a good search takes practice and you'll probably get the best results if you consult a librarian.

What's Next?

Once you have a final search set up and ready to go, you are ready to run it. But, after you run it, may need to make some adjustment to it to **refine** it. We'll cover those techniques in the next chapter.

CHAPTER 26
REFINING SEARCHES

This chapter is all about *refining* your search to make it just right. You'll learn how to react to your results and then execute strategies to tighten or expand your search if needed. Plus, you'll gain a better of idea of what constitutes a reasonable number of results. These are the last skills you need to be an adept searcher. After that, all you need is practice!

Taking Stock of Your Initial Results

We're skipping over step 5 in the search process because you already learned how to choose databases to search in Chapter 23. That means we're ready for step 6 (see Table 26.1) where you **run your search**, which just means entering your keywords/synonyms into the databases you've chosen and then clicking the "search" button.

Table 26.1

Steps in the search process	*Covered in*
1. Identify the fewest, most essential keywords that represent your topic.	Chapter 24
2. Gather synonyms for your keywords.	Chapter 25
3. Modify your keywords/synonyms with truncation and quotation marks.	Chapter 25
4. Set up a Boolean search.	Chapter 25
5. Choose databases to search.	Chapter 23
6. Run your search, then refine it as needed.	This chapter

Next, you'll want to take stock of your results. The first thing to note is how they're sorted. Nowadays most databases deliver results **sorted by relevance**, with the most relevant sources listed first. How do databases know which sources are the most relevant? They don't. They just use algorithms to identify database records that have the most matches to your keywords, and they put those records at the top of your results list. (The algorithms are a bit more complex than that, but that's the gist of it.) This method of ranking relevance is a bit crude, but it's still useful; if you scan through the first page or two of your results and see at least a few sources that look potentially relevant, then that's a good sign your search is on the right track. Note that most databases provide several options for sorting your results, so if you prefer to see them sorted by date published, or in some other order, then you can make that adjustment.

The next thing to note is how many results your search retrieved. Suppose you're searching Web of Science and the search you entered retrieved over 10,000 results. In that case, you can still scan through the first few pages of your results to see if there's anything relevant, but you may want to work on tightening your search to get more targeted results. On the other hand, if you only got a few results—such as fewer than twenty—then you should consider expanding your search to collect more results. You'll learn techniques for tightening and expanding shortly.

What's a Reasonable Number of Results?

The answer to this question depends on the circumstances. If you're working on a college assignment and you only need to find a few sources, and you've run a search that turned up some good-enough sources, then you're done. You've found what you need.

It's in situations where you're trying to run a *comprehensive search* on a topic to find *all* relevant publications that you might wonder what's a reasonable number of results to aim for. As mentioned in previous chapters, situations where you need to search comprehensively are likely to arise when you embark on your chosen career as a scientist or engineer. So, let's prepare you for that!

Suppose you're running a comprehensive search for articles about knee replacement rehabilitation, and you get 334 results from the database PubMed. Most students are not happy with the idea of getting that many results because they think it'll take forever to look through all of them to find the best articles. However, this is actually a very reasonable number of results. Why? For one thing, you may end up filtering your results, and this can quickly reduce them to a more manageable number.

For example, perhaps you're only interested in looking at the most current sources about knee replacement rehabilitation, in which case you might impose a **publication date** filter so that you're only viewing sources published within the last five years. Plus, you might only be interested in viewing certain *types* of sources, such as journal articles describing clinical trials. If so, you can further filter your results by **source type** (also called **document type**). Applying filters like these (publication date and source/document type) can reduce 334 results to fifty-seven results in just a few clicks. If you need more information about filters, consult Chapter 22.

Another reason you needn't balk at a few hundred results is that you may not need to look through all of them anyway. Assuming your results are sorted by relevance, you might glance through the first several pages of results to identify those that seem useful for your research project. As you get more pages in, you may start to see results that look less and less relevant—until you aren't seeing any that look promising. So, you might decide to stop at that point without looking through the rest.

But, the most important reason not to be skittish about getting several hundred results is that it's often fairly easy to look through all of them, as you'll see in the next section.

How to Approach Scanning through Several Hundred Results

Let's look at how scanning through several hundred results might work, and why it's not that onerous. Suppose you're searching a database for sources about horses from the Eocene epoch, and you entered this search into GeoRef using two stacked search boxes:

> **eohippus** OR **hyracotherium** OR **equoid** OR **equid*** OR **horse** OR **equine OR perrisodactyl**
>
> AND **Eocene**

This search retrieves 303 results and you decide to look through all of them because you're working on a project where you really need to be thorough and find *everything* that's relevant to this topic. Assuming your results are sorted by relevance, the first few pages will likely have the most relevant results, as you learned earlier in this chapter. So, you'll want to go through those pages carefully and note which records are worth saving to look at in more detail later. (You'll learn how to mark and save records in the next chapter.) But, as you get more pages in, your results will probably be less and less relevant and it'll take just a few seconds to read the titles to determine that they're not worth saving.

In other words, as you get more pages into your results, it usually gets quicker and quicker to scan through a page of results and choose just those that interest you (if there are any). In fact, going through a few hundred results to identify just those you want to save usually takes about twenty minutes or so—much less time than most people expect. So, if you want to search as thoroughly as possible and not miss anything, it's better to just look through *all* your results. Of course, if you have thousands of results, then you'll need to strategically tighten your search so you have a more manageable number to look through. So, let's talk about how to do that.

Refining Your Search to Get Fewer, More Targeted Results

If your search retrieves several thousand results rather than a few hundred, then there are several techniques to tighten things up to get a more targeted set of results.

1 Apply Filters

As mentioned earlier, you can apply **filters** to your results (also called **limits**, **refinements**, or **search options**). Filters are usually displayed to the left of your results. The filters people tend to use the most are publication date, source/document type, and language, but depending on the database you're using, there may be other filters that would be extremely useful to employ. For more information about filters, see Chapter 22.

2 Utilize Phrase Searching (When Appropriate)

In the last chapter you saw how **phrase searching** can eliminate lots of irrelevant results. Recall when we searched Web of Science for **forever chemicals** we got 263 results. But when we changed this to the phrase search you see below, we got ninety-eight results, and those results were much more on-topic:

> **"forever chemicals"** OR **"forever chemical"**

3 Restrict Some/All Keywords to Certain Fields

Another way to tighten your search is to consider whether there are some keywords you could strategically restrict to a particular field. Here's an example—suppose you wanted to quickly find some good articles about developing lesson plans for chemistry or biochemistry education. Figure 26.1 shows one way you could set that search up. Notice that the keywords in the top box have been restricted to the **Title** field. This tells the database that any results it retrieves must have those keywords in the title of the source.

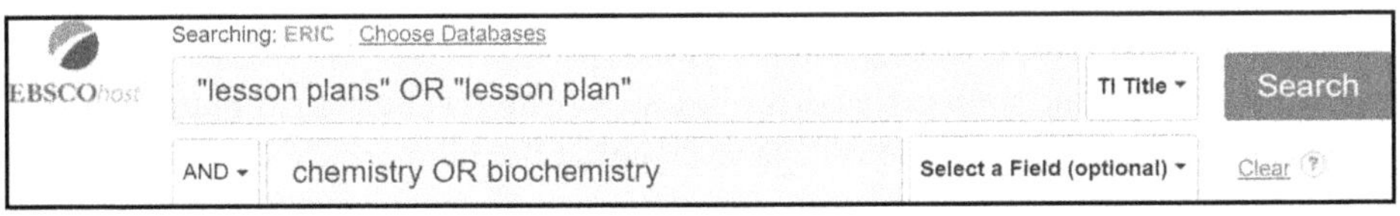

Figure 26.1

Here's the rationale for doing this: if an author has written a source that's primarily about the creation of lesson plans, then it's likely they would have given their source a title that includes the phrase **"lesson plans"** OR "**lesson plan"** like you see in the database results shown in Figure 26.2.

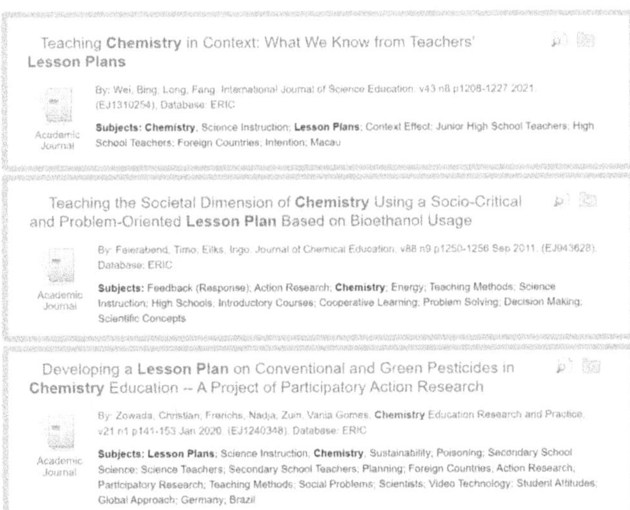

Figure 26.2

So, when you force a database to find keywords in the title of a source, you'll increase your chances of retrieving results that are highly relevant. You'll also get fewer results because you've told the database it can't retrieve records with the phrases **"lesson plans"** OR **"lesson plan"** just anywhere in a full record—rather, it must only retrieve records with those words in the *title*.

Be aware, though, restricting keywords to the title field means you *might* miss out on relevant results. For example, suppose an author wrote an article about creating chemistry lesson plans but entitled it: *Tried and true techniques for effectively teaching chemistry*. Your search won't find that article because it doesn't have either of the phrases **"lesson plans"** OR **"lesson plan"** in the title. Nevertheless, restricting keywords to the title can help you zero in on a smaller set of relevant articles, and sometimes that's extremely useful!

4 Narrow Your Topic by Adding Another Dimension to It

If you're getting way too many results, and it's possible to reconsider your topic to make it narrower, then that's another way to get a tighter, more focused set of results. Narrowing your topic usually means adding another dimension to your search. For example, suppose you want to find sources that describe lesson plans for chemistry or biochemistry education—but specifically at the college level, not at the high-school level. When you add in keywords that represent this new dimension—such as the keyword **college** along with a couple of synonyms that come to mind (**university** or **undergraduate**)—the search will get fewer results because the topic is narrower. The new dimension you added creates a more focused search. Here's how you might enter that search into three stacked search boxes in a database:

> **"lesson plans"** OR **"lesson plan"**
>
> AND **chemistry** OR **biochemistry**
>
> AND **college** OR **university** OR **undergraduate**

For more tips on narrowing topics, see Chapter 19.

Expanding Your Search to Find More Results

Now let's suppose you ran a search and got very few results. How could you expand it to ensure you're finding everything that's relevant? Here are some techniques.

1 Reduce the Number of Keywords You're Searching

Entering a search as a sentence is the most common reason for getting too few results. Recall that when you enter keywords into a database, the database assumes you want to connect those words with the Boolean operator **AND**—which is the same as telling the database "You must retrieve records with *all* of these keywords." So, the more words you put in the search box, the more the database has to work to find and retrieve records that have every single one of those words. To loosen your search and retrieve more results, search the fewest, most essential keywords, like you learned to do in Chapter 24.

2 Find and Add More Synonyms for Your Keywords

As you learned in Chapter 25, finding synonyms for your keywords, and then setting up a Boolean search, is an excellent way to expand your search and ensure you get a good, comprehensive set of results. As an example, consider this topic:

- **What factors contribute to death from eating disorders?**

Of the two searches in Table 26.2, the second one—where the word **mortality** is added as a synonym for **death**—will get *more* results. When you add synonyms and connect them with the Boolean operator **OR**, you're giving the database more choices of words that are ok for it to find and retrieve, thus expanding your search.

Table 26.2

Search 1	*Search 2*
death AND **"eating disorders"** OR **"eating disorder"**	**death** OR **mortality** AND **"eating disorders"** OR **"eating disorder"**

In case you need a mini-review of Boolean operators, recall that keywords connected by **AND** will *restrict* your search, whereas adding synonyms for your keywords and connecting them with **OR** will *expand* your search. For more information about finding and adding synonyms to your search, consult Chapter 25.

3 When Relevant, Truncate Your Keywords

If you are using keywords that could have several different word endings, then truncate them. This will help you retrieve more results. Refer to Chapter 25 for more information about truncation.

4 Broaden Your Topic by Reducing the Number of Dimensions It Has

If you are getting too few results, and it's possible to reconsider your topic and make it broader, that could help you get more results. You saw an example of this back in Chapter 19 when we looked at this topic:

- **Does dark sky community status boost tourism in Flagstaff, Arizona?**

For that topic, we might set up a search in a library database using three stacked boxes like this:

> **"dark sky"** OR **"dark skies"**
>
> AND **tourists OR tourism**
>
> AND **Flagstaff**

So there are three main concepts or dimensions to this search, and we've joined them together with the Boolean operator AND. The more dimensions you connect with AND, the narrower your search will be. If you remove a dimension, you will broaden your search. For this search, it makes sense to remove the geographic dimension (Flagstaff) and instead look for articles about how dark sky communities increase tourism, regardless of the location. So, you could restate the topic as follows:

- **Do efforts to preserve dark skies promote tourism?**

And search it like this instead:

> **"dark sky"** OR **"dark skies"**
>
> AND **tourists OR tourism**

What If You Can Only Find a Few Good Sources on Your Topic?

Suppose you've been searching for sources on a topic, but so far you've only found one or two that are relevant, even after trying the techniques listed in the section above. For example, suppose you're looking for journal articles about **blindsight** and specifically the different ways it manifests, and the only really good article you've found so far is this one:

> Danckert, J. & Rossetti, Y. (2005). Blindsight in action: what can the different subtypes of blindsight tell us about the control of visually guided actions? *Neuroscience and Biobehavioral Reviews*, *29*(7): 1035–1046.

In that case, turn to these techniques:

1 Try "More Like This" Searching

Many databases provide some form of "More like this" searching, which can sometimes be effective for finding additional, similar articles. To use this technique, simply copy and paste the title of your source (for example, **Blindsight in action: what can the different sub-types of blindsight tell us about the control of visually guided actions**) into some databases to see if you can find a record for that article. If you do, open the full record and then look for links that will lead you to more articles like that one. For example, in Google Scholar, you'd click **"Related Articles"** (see Figure 26.3).

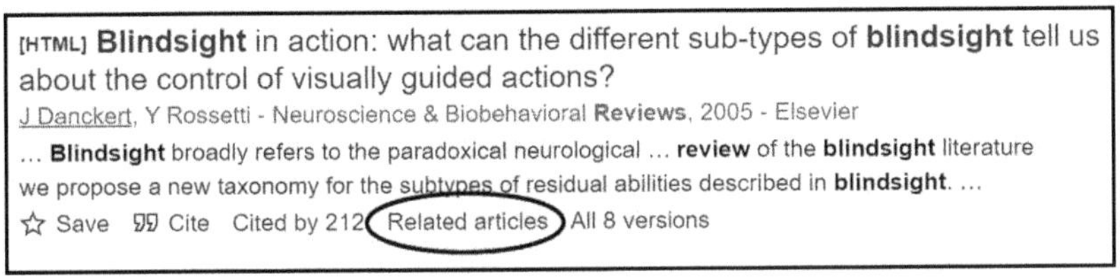

Figure 26.3

However, note the link will be worded differently and appear in a different spot in other databases. In PubMed the link is called **Similar articles**, and in Web of Science look under the heading **View Related Articles.**

These databases use various algorithms to identify and deliver similar articles, but the algorithms don't always get it right; sometimes you find good stuff, other times it's a bust. Still, it's worth trying.

2 Try Searching for Authors

Note the authors of your source and try searching them to see if they've written and published more on the same topic. Most databases offer a way to search for sources by a particular author. In Web of Science, for example, you can select **Author** from the drop-down search menu (see Figure 26.4).

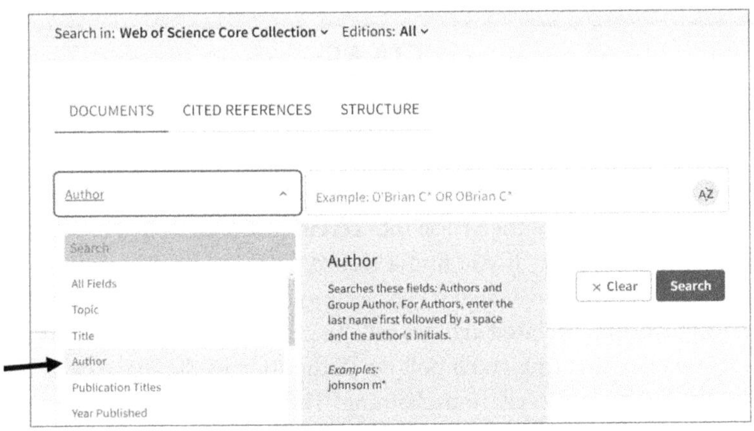

Figure 26.4

3 Explore Citation-Chaining

Citation-chaining is another technique for finding additional, relevant sources. This technique involves taking a source (like the Danckert and Rosetti article about blindsight) and looking at the sources it cites. They'll be listed at the end of the article in a section called **References Cited** or something similar. Scan through those cited sources to see if you come across additional relevant sources that would be useful to you.

Then, look at the sources that have cited the Danckert and Rosetti article to see if any of *those* sources look relevant to your topic. The idea here is that you're likely to find more sources on the same topic by looking at a source's **cited** references as well as the sources that **cite** it; this strategy is illustrated in Figure 26.5.

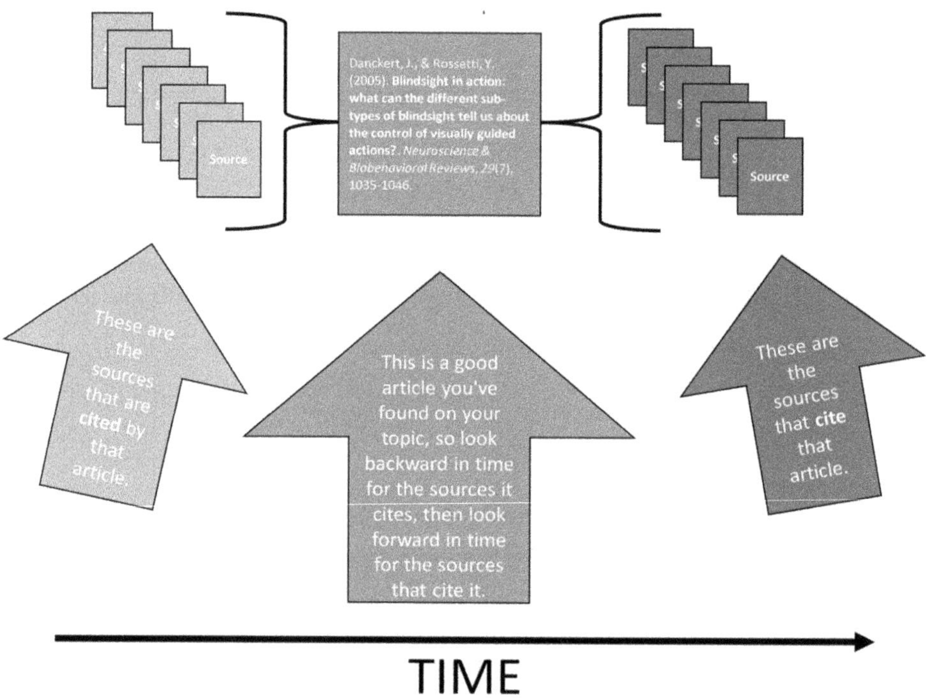

Figure 26.5

But how do you find the sources that have cited the Danckert and Rosetti article? Just copy and paste the title of the article into certain databases, such as Web of Science, Google Scholar, PubMed, etc. If you find a record for that source, you'll see options to view the publications that have cited it. For example, take another look at Figure 26.3. Notice that just before the **Related articles** link, there is a link called **Cited by**, along with a number. If you click that link you'll pull up all the sources that have cited that article. So, now you know how to do citation-chaining! This technique for finding articles can be extremely useful so don't forget about it.

4 Use AI Literature Searching Tools

You learned in Chapter 20 that there are several new artificial intelligence (AI)-driven literature search tools such as Scite.ai, Elicit.org, and ResearchRabbit.ai. While these tools are not ideal for conducting robust, comprehensive searches on a topic because of flaws such as the black box problem and the Matthew effect (see Chapter 20 for details), you can leverage them to do all the functions described above: author searching, more-like-this searching, and citation-chaining.

To sum up, what do science and engineering students need to know about refining searches?

1. Take stock of your initial search results to note how they're sorted and whether you're finding at least *some* relevant sources; then note how many results were retrieved and whether it might be useful to tighten or expand your search.
2. If you get too many results, consult the various techniques listed in this chapter to tighten up your search.
3. If you get too few results, consult the various techniques listed in this chapter for expanding your search.

What's Next?

You've finished **Part VI** in which you learned all about searching databases to find sources. Now you'll move on to **Part VII** which will explain how to **manage source records** and get access to **full text**.

PART VII

ORGANIZING SOURCES AND GETTING FULL TEXT

CHAPTER 27
MANAGING SOURCE RECORDS

When you run searches in databases and find records for promising sources, you'll want to save, export, and organize them. This chapter will explain how to do those things, including how to use reference managers to store and organize the sources you've found. If you're managing sources for a long-term project, you may need to keep them up to date so this chapter will also explain how to set up search alerts to capture the latest published articles on your topic.

Selecting, Saving, and Exporting Records

There are several ways to select and export records from a database, and one way *not* to do it. Let's start with what you shouldn't do. If you have one or more database results you want to save, *do not* go to the top of the page and copy the URL for the database page where those records appear. While that URL might lead you back to those records while your database session is still running, it probably won't work if you try to click on it the next day. Often these URLs are only associated with your current database session and will expire when your session is over. So instead, use one the four techniques listed below to save your database records. There's also a fifth way to save records using a reference manager and we'll discuss how to do that later in this chapter.

1 Export All Records Found

If you want to save and export *all* the results your search retrieved, most databases provide ways to do that (as long as you don't have thousands of results). While the exact process varies from database to database, typically you look for options labeled **save** or **export.** Selecting that option will result in a prompt asking any or all of the following:

1. How do you want to export the records—do you want to email them to yourself, or do you want them in a Word document, or an Excel spreadsheet, or formatted to import into reference management software?
2. What span of records do you want to export—do you want all of them, or just the first fifty, or some other range?
3. What bibliographic details do you want to export—do you want just the basic information needed for citing the source, or do you also want the abstract, the associated subject terms, etc.?

For example, in Figure 27.1 you see a search in Web of Science for sources about toad toxin biosynthesis. This search resulted in eleven records. To export all eleven, click the **export** button, then select how you want the records exported from the drop-down menu.

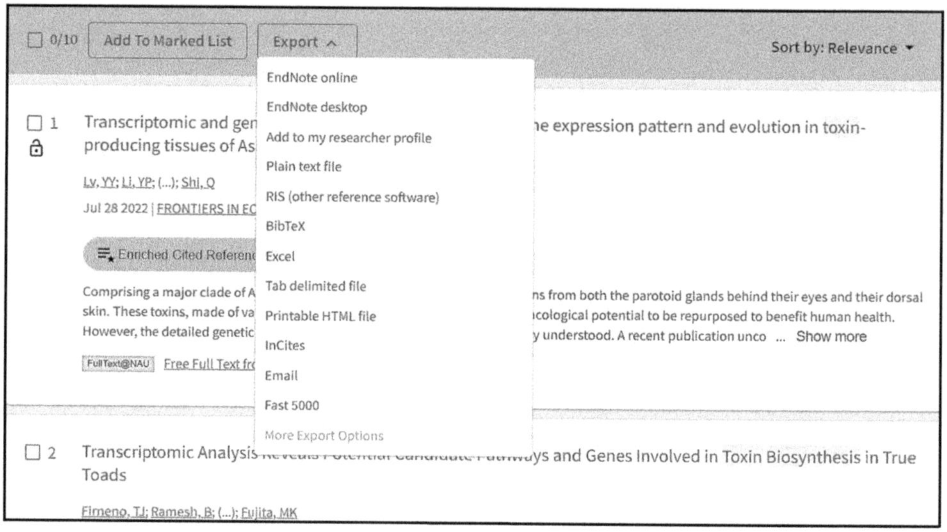

Figure 27.1

Let's suppose you chose to export to an Excel file. Once you make that selection, you'll be asked to specify which span of records to export and what bibliographic information you want to export. For each database this process looks and works a bit differently, but consult Figure 27.2 to see how the process looks in Web of Science. Again, for each database this process looks and works a bit differently, but in Web of Science, here's what you'd see.

2 Mark and Export Specific Records

Many databases allow you to mark and save just the records that interest you, and then export those records. To mark records, there's usually a box to the left of each brief record that you can check off in order to mark and save it. In some databases, the option to mark records is on the right side of the record and appears as a folder you can click, rather than a box. Because the exact procedure for marking, saving, and exporting records varies

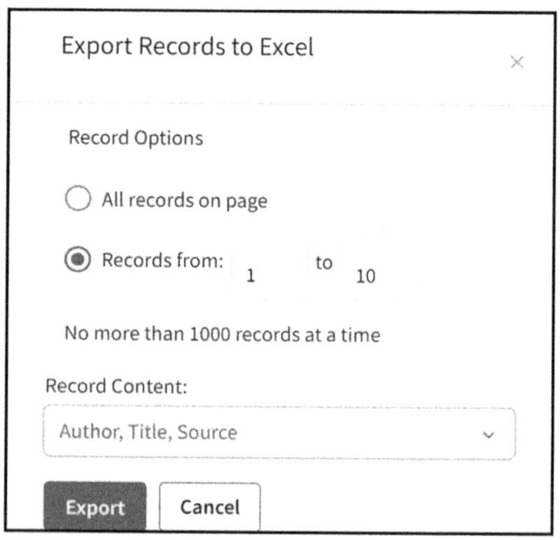

Figure 27.2

from database to database you may need to ask a librarian for help, or you can try looking up instructions yourself using Google searches such as:

- **[database name] saving records**
- **[database name] (export OR exporting)**

Note that if your database search resulted in a lot of records to sort through, you should not start marking records, then walk away and have lunch or take a nap because when databases detect inactivity they'll end your session, and you'll lose all the work you did. Instead, if you need to take a break from marking records, then export what you've already marked, and note where you stopped. When you're ready to continue, you'll need to reproduce the same search, then page forward in your results to where you left off and commence marking again.

3 Save Persistent Links to Records

In the introduction to this section, you learned *not* to copy the URL for the page where your database records appear because it may not lead you back to those records. However, some databases *do* offer a **persistent link**, also called a **permalink**, that you can copy and save to get back to a specific record. In Figure 27.3 you see a permalink that's offered in the full record for a source. If you were to click that permalink icon, a link would be presented to you that you could copy and save to get back to this source record at any time.

4 Copy and Paste Records into a Word Document

Of course, if you only need to save a few records and you want to keep it simple, you can also just scroll over a database record to highlight and copy it, then paste it into a Word document to save it.

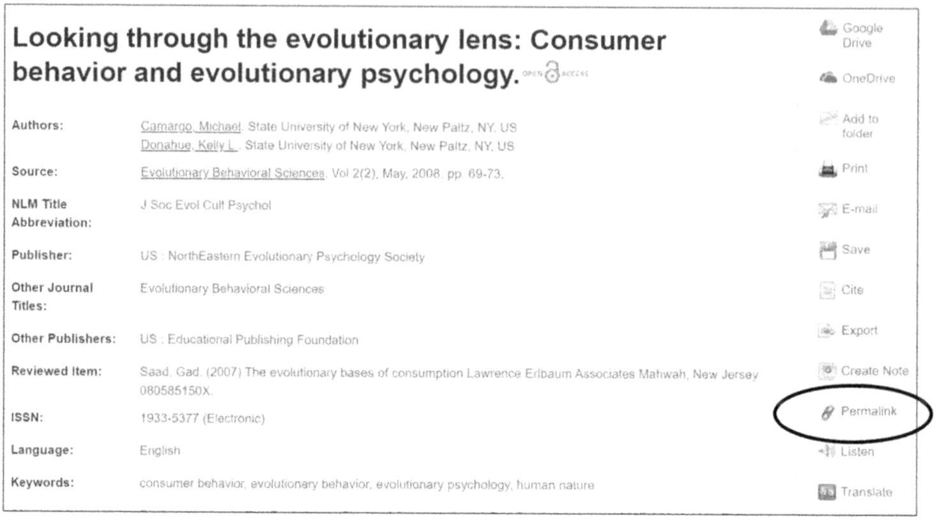

Figure 27.3

Store and Organize Records Using a Spreadsheet

If you're exporting lots of records from a database, you'll probably want to import them to a spreadsheet or a reference manager to store and organize them. Let's talk about spreadsheets first. A **spreadsheet** is a grid that allows you to input data into columns and rows. You can also sort and manipulate those data in various ways. One advantage to using a spreadsheet is that you can create your own columns that track characteristics of each source, like you see in the template in Table 27.1.

Table 27.1

Source citation	*Research question*	*Number of subjects*	*Study design*	*Type of treatment*	*Year data were gathered*	*Conclusion*

This organizational method can be especially useful if you're writing a review article and you want to document certain details about each source you found so you can identify trends, discrepancies, etc. (See Chapter 8 for more information about review articles.)

The spreadsheet you see above is simplified, by the way. If you were to export citations from a database to a spreadsheet, it would split all the bibliographic information about each source into its own column, like you see in Table 27.2.

Table 27.2

Author	*Article title*	*Source Title*	*Publication Year*	*Volume*	*Issue*	*Start page*	*End page*
Wang, Y.	Entanglement in quantum process algebra	International Journal of Theoretical Physics	2019	58	11	3611	3626
Del Cima, O. M., et al.	Magnetic shielding of quantum entanglement states	Quantum Studies—Mathematics and Foundations	2019	6	2	141	150
Fognini, A., et al.	Dephasing free photon entanglement with a quantum dot	ACS Photonics	2019	6		1656	1663

Plus, your export might include lots of other fields as well, such as author addresses, ISSN numbers, DOIs, etc. You can just hide or delete the columns of data you don't need, and then add in your own columns for the additional data you want to record, such as research question, study design, etc.

Store and Organize Records Using a Reference Manager

Reference managers allow you to create your own personal database of source records that are of interest to you. You can store journal article citations, book citations, website citations, or citations for any other type of source. Popular examples include subscription-based reference managers like RefWorks and EndNote or freely available options like Mendeley and Zotero. You may hear reference managers called a variety of different names including:

- **reference managers**
- **bibliographic managers**
- **citation managers**
- **reference management software**
- **bibliographic management software**
- **citation management software**
- **reference management tools**
- **bibliographic management tools**
- **citation management tools**

There are multiple ways to populate your reference manager with source records. One way is to export a file of records from a database, then import that file of records to your reference manager. Another way is to auto-capture records using what's called an **extension**, **plug-in**, or **add-on**. Suppose, for example, you're using the reference manager Zotero, which offers an extension called Zotero Connector. Zotero Connector works with the Chrome browser to sense sources you find in databases. It then captures relevant bibliographic information about those sources so you can easily import them to Zotero.

Reference manager extensions such as Zotero Connector can also sense and capture bibliographic data from a webpage, an online newspaper, magazine, or journal article—as well as a PDF, an online book, etc. However, you should be aware that sometimes the data captured are incomplete. If you're trying to capture the bibliographic data for a source and your extension tool doesn't gather everything you need—for example, perhaps it misses the publication date—then once you've imported the record to your reference manager you may need to manually edit it and add in the missing information.

You can also manually add records to a reference manager. Just look for an option to add a new source, then select the type of source you want to add, such as a journal article. You'll then be presented with a form where you can add in the details you have about that source.

Reference managers also allow you to organize records into different folders that represent different projects you are working on. So if you think you are going to be working on a project over multiple semesters while in college or graduate school, then it a makes a lot of sense to start storing sources related to that project in a reference manager. It'll keep you a lot more organized!

What Else Can Reference Managers Do?

Some reference managers allow you to download and attach PDF files of journal articles or conference papers to your source records, and some will automatically download available PDFs when you export records using the reference manager's extension feature. You can also add notes or tags to records. For example, if you're working on a project about disease transmission education campaigns around the world, you could tag your sources by country so you could search and sort your sources by a particular country.

Another feature these tools offer is duplicate removal. If you import records from multiple databases, you may end up with duplicate records for the same source. If so, your reference manager can detect and remove these duplicates. And, if you're collaborating with others on a project, some reference managers will allow you to share and edit folders of stored records with your collaborators.

Reference managers can also output all your stored records, or just records that you specify, into formatted citations to create an instant bibliography. Most reference managers have lists of different citation styles for you to choose from to format your bibliography. You'll learn more about this in Chapter 31.

Choosing a Reference Manager and Getting Help Using It

If you want to start using a reference manager, then you have lots of options to select from. You can use one of tools mentioned earlier in this chapter (RefWorks, EndNote, Mendeley, or Zotero) or you can explore other options with these Google searches:

- **free (reference OR citation OR bibliographic) (software OR manager OR management OR tools)**
- **lists (reference OR citation OR bibliographic) (software OR manager OR management OR tools) lists**
- **(choose OR choosing) (reference OR citation OR bibliographic) (software OR manager OR management OR tools)**

You might also check with staff at your campus library to see if there are reference managers they recommend.

If you need help learning how to use a reference manager, or learning how to perform a particular function, then check to see if your campus library provides technical support. Alternatively, you can search Google for the name of your reference manager along with the function you want to learn, and you'll likely find sites with written instructions and/or videos explaining what to do. For example, this Google search will result in helpful sites for learning how to import citations to Zotero:

- **Zotero (import OR importing)**

Using "Alerts" to Keep up to Date

Suppose you're working on an ongoing project that requires staying up to date with the latest research. In that case, you might want to save your database searches and set up **alerts** so that anytime a database adds a new record that fits your search criteria, you'll be notified, and then you can download that new record.

Many databases offer an option to save your search and set up an alert, but to do so, you'll need to sign up for a free personal account with the database, so it can store your searches. Usually there is a link to set up an account on the database's home page. Once you've established an account, then you can set up alerts. In Web of Science, for example, once you've set up a search and run it, you'll see an option to create an alert for that search as shown in Figure 27.4. You can log into your account at any time to view or edit your saved searches, add or delete alerts, etc.

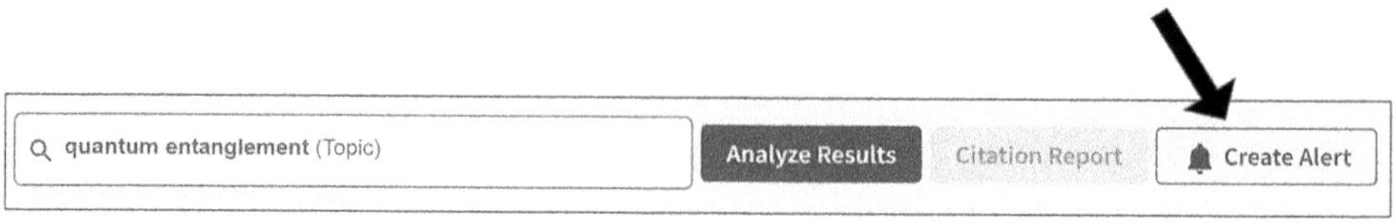

Figure 27.4

To sum up, what should science and engineering students know about managing source records?

1. Most databases offer various ways to save and export either a *batch* of records or *selected* records into a Word document, Excel spreadsheet, reference manager, etc.
2. Reference managers allow you to create your own personal database of sources, plus many offer features allowing you to add PDFs, add notes to source records, etc.
3. Many databases offer a way to set up alerts so you can keep up with the latest sources published on topics of interest to you.

What's Next?

Now that you've learned how to export and manage your source records, you'll want to get the **full text** for them so you can read them in their entirety. The next chapter will cover that.

CHAPTER 28
ACCESSING FULL TEXT

"Full text" is a word that's come to mean the entire article, conference paper, book, or other text-based source. Up until now you've learned how to find, store, and manage *records* for sources, rather than interacting with the source itself in its full text form. So, this chapter will explain various ways to gather and store the full text, and what to do if you run into difficulties tracking down the full text.

Getting to Full Text from a Database Search

If you're searching a database to find sources, such as one of the databases available on your campus library's website, each record in your results will usually provide a way to get to the full text. Just look for links, buttons, or icons with wording like this:

- **View full text**
- **HTML full text**
- **Read online**
- **Full text from publisher**
- **Download PDF**
- **PDF**
- **Fulltext@MyCampus**
- **Online access**

Alternatively, the record may have a link or button with this type of wording:

- **Request from Interlibrary Loan**
- **Get from Document Delivery Service**

Regardless of the exact wording, these sorts of links, buttons, and icons usually show up just under each brief record. To see examples of how this looks, refer to Figure 21.1. However, in Google Scholar you'll see full-text links to the right of your results. And in some databases like PubMed, you'll need to open the full record to see the links to full text. What happens when you click on one of these links, buttons, or icons? Any of the following could happen:

1. You might immediately be presented with a full-text HTML or PDF version of the source that you can read and/or download.

2. You might be sent to an intermediate page where you'll have to click another link, button, or icon to continue proceeding to the full text. Often it's a link that designates a particular collection of content; for example, it might be worded: **Full-text availability from: Elsevier Science Direct Journals**.
3. You might be sent to a page that presents a link to your library's document delivery/interlibrary loan service. If you recall from previous chapters, this service helps you obtain sources your library doesn't offer immediate access to. Some libraries call the service "document delivery" and others call it "interlibrary loan." Note that if you're requesting an article or paper through this service, it may take a few hours or even a few days to arrive. If you're requesting a book, it may take a couple of weeks to arrive. Regardless of the type of source you're requesting, when it arrives, you'll be notified by email and provided with instructions for how to download the electronic document, or how to pick up the physical item from the library.

   Note that if you're looking for the full text of a source and you end up getting directed to your library's document delivery/interlibrary loan service, then before placing your request, first look to see if you can find a free copy of the source online, as that'll be a quicker way to get it! Just open a new window or tab in your browser and try running a search on the title of the source in Google to see if you can find a free copy. If you can't find a free copy, then go ahead and initiate a request for the source using your library's document delivery/interlibrary loan service.
4. You may run into any number of different problems where you're unable to navigate to the full text of the source. Perhaps you encounter an error page, or you land on a page that doesn't seem to offer any options for getting the full text. If that happens, first try running a Google search on the title of the source to see if you can find a copy of it for free somewhere on the internet. If that doesn't work, then see the section below on dealing with full text access problems.

What If a Database Record Has No Links to Full Text?

You'll probably encounter situations where you're searching your library's catalog or discovery tool (see Chapter 20 for more information about these databases) and you see records that have no links, buttons, or icons leading to full text. This could be because the record represents an item in your library's physical (print) collections, rather than an electronic document. If that's the case, you should see a location and call number within the record that will help you find the item in the library, like in the example shown in Figure 3.3.

Or, if you're searching a database and see no way to get to the full text of the source, or any way to locate a physical copy of the source within the library, then try searching both Google and Google Scholar for the title of the source to see if you can find and access it. If that doesn't work, see the section below on dealing with full text access problems.

Tracking Down Full Text for a Source You've Encountered Elsewhere than a Database

Let's suppose instead of searching a database for sources, you happen to be reading this article and you want to track down one of the references cited in it:

> Gorman, J. C., et al. (2021). Diverse processing underlying frequency integration in midbrain neurons of barn owls. *PLoS Computational Biology, 17*(11): e1009569.

Figure 28.1 shows the specific publication you want to track down, as it appears in that article's **References** section.

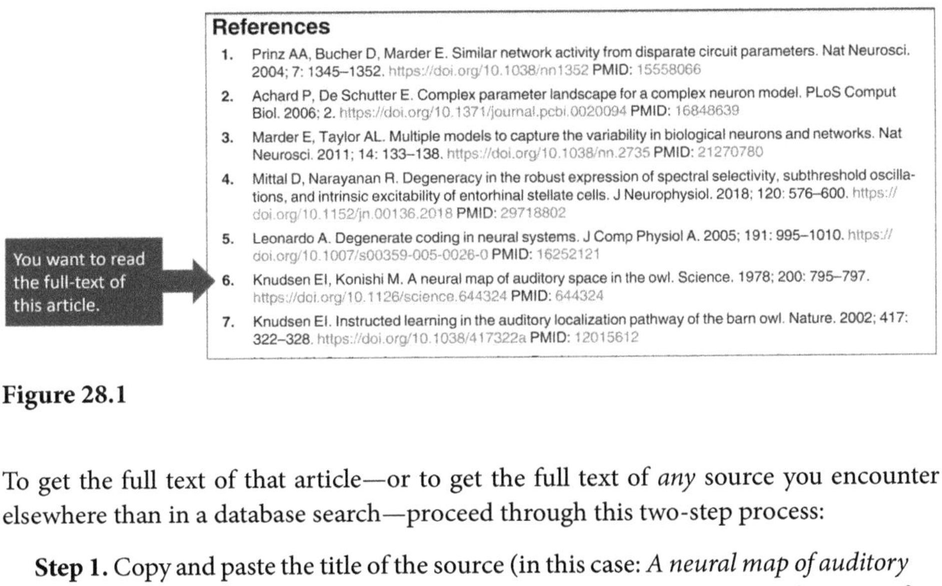

References

1. Prinz AA, Bucher D, Marder E. Similar network activity from disparate circuit parameters. Nat Neurosci. 2004; 7: 1345–1352. https://doi.org/10.1038/nn1352 PMID: 15558066
2. Achard P, De Schutter E. Complex parameter landscape for a complex neuron model. PLoS Comput Biol. 2006; 2. https://doi.org/10.1371/journal.pcbi.0020094 PMID: 16848639
3. Marder E, Taylor AL. Multiple models to capture the variability in biological neurons and networks. Nat Neurosci. 2011; 14: 133–138. https://doi.org/10.1038/nn.2735 PMID: 21270780
4. Mittal D, Narayanan R. Degeneracy in the robust expression of spectral selectivity, subthreshold oscillations, and intrinsic excitability of entorhinal stellate cells. J Neurophysiol. 2018; 120: 576–600. https://doi.org/10.1152/jn.00136.2018 PMID: 29718802
5. Leonardo A. Degenerate coding in neural systems. J Comp Physiol A. 2005; 191: 995–1010. https://doi.org/10.1007/s00359-005-0026-0 PMID: 16252121
6. Knudsen EI, Konishi M. A neural map of auditory space in the owl. Science. 1978; 200: 795–797. https://doi.org/10.1126/science.644324 PMID: 644324
7. Knudsen EI. Instructed learning in the auditory localization pathway of the barn owl. Nature. 2002; 417: 322–328. https://doi.org/10.1038/417322a PMID: 12015612

Figure 28.1

To get the full text of that article—or to get the full text of *any* source you encounter elsewhere than in a database search—proceed through this two-step process:

Step 1. Copy and paste the title of the source (in this case: *A neural map of auditory space in the owl*) into your library's discovery tool, if your library has this type of database. (If not, proceed to Step 2.) Hopefully you'll find a record for the source at or near the top of your results, and the record will have a link leading you to the full text. If not, go on to Step 2.

Step 2. Try pasting the title of the source into both Google and Google Scholar—that's right, check both. If you don't find it, or you find it but can't access the full text, then refer to the section below on dealing with full text access problems.

Dealing with Full Text Access Problems

If you're having trouble getting access to the full text for a source, and you've already tried conducting Google and Google Scholar searches to find it for free on the internet but with no luck, then ask library staff for help, or go ahead and initiate a request for the source through your library's document delivery/interlibrary loan service. You can usually find a link to this service somewhere on your campus library's home page. If you can't find it, then ask for help.

Configuring Google Scholar to Optimize Full Text Availability

Just so you're aware, there's a way to configure Google Scholar to detect your library's subscriptions to full text content. To set this up, refer to Figure 28.2 and follow these steps:

1. Click the three stacked bars in the upper left of the Google Scholar search page.
2. Select **Settings**.
3. Select **Library links**.
4. Enter and search the name of the campus that you're affiliated with.
5. Click the boxes next to any relevant results that appear.
6. **Save** your settings.

If these steps don't work for you, then check with your campus library for instructions.

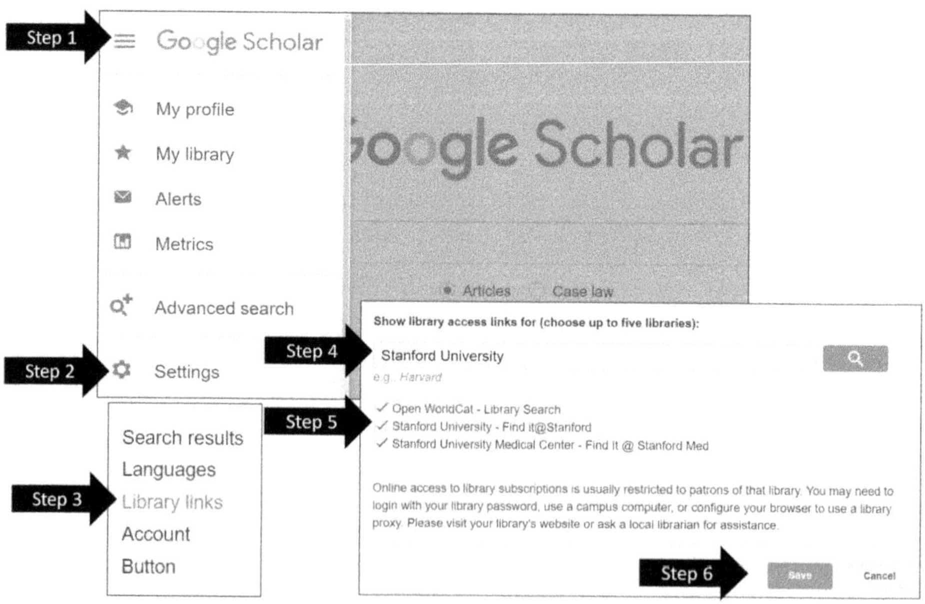

Figure 28.2

Storing Your Full Text

There are several options for storing PDFs of your full text sources. You can create storage files on your computer, or on an external drive, or in a cloud-space like Google Drive, Dropbox, etc. Just make sure that if you're storing files on your computer hard drive, you have them backed up elsewhere in case your computer fails. Or, if you're using a reference manager, you may be able to store PDF files within your reference manager if it offers this capability.

To sum up, what should science and engineering students know about accessing full text?

1. To get the full text of a source you found while searching a database, look for links, buttons, or icons within the database record that look like they'll lead to full text.
2. If you don't see any options to get to full text, then check if the database record indicates the source is located on one of the shelves in the library.
3. If there's a source for which you need the full text and you happened upon it in some other way than searching a database, then look it up by title in your library's discovery tool. If your library doesn't have a discovery tool or if you don't find the full text there, then check both Google and Google Scholar.
4. Whenever you run into problems getting full text, use your library's document delivery/interlibrary loan service to request the source, or simply ask for help!

What's Next?

You're finished with **Part VII**, so now you know how to better manage your sources, plus you learned various ways to get the full text of your sources! Next up is **Part VIII**, which is all about **citing** your sources. You'll learn why it's important to cite, how to incorporate citations into your writing, and how to format citations.

PART VIII
CITING YOUR SOURCES

CHAPTER 29
THE IMPORTANCE OF CITING

Citing sources signals your conscientiousness and credibility as a researcher and a writer, so it's pretty important! In this chapter, you'll learn what citing is, where you need to cite in your writing, and how citing helps contextualize and substantiate the content you're communicating to readers. This chapter also explains how citing functions as an acknowledgment of the hard work and ideas you've borrowed from others. Failure to properly acknowledge and cite others' work is an act of plagiarism, a serious offense that this chapter will explain, so you can be sure to avoid it!

What It Means to Cite Sources

When you write, sometimes you need to consult and use content from outside sources. That means you *must cite* those sources. To **cite** means to provide your readers with a **citation** (also known as a **reference**) showing you used information from an outside source. A citation is an abbreviated way to refer to a source; it includes just enough information so that others could track down the source. You've seen examples of citations throughout this book, and Figure 29.1 shows you a journal article citation with all the parts identified.

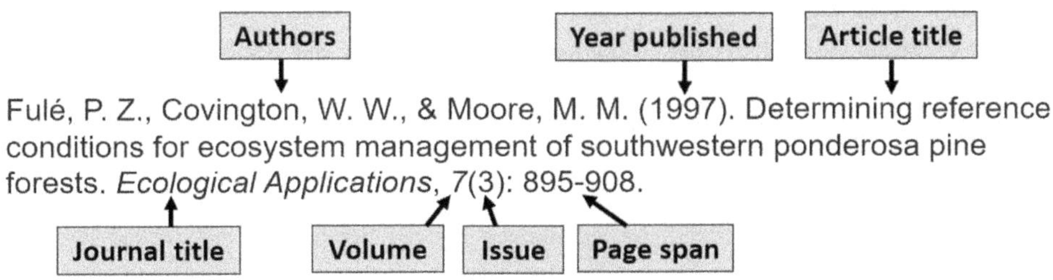

Figure 29.1

So, citing conveys to your reader that you used an outside source, and it provides your reader with relevant details for finding and consulting that source for themselves, should they want to. Note that the order and formatting of elements in a citation are dictated by the particular **citation style guide** you are using. We'll talk about citation style guides in Chapter 31.

Where Should You Cite in Your Writing?

Typically you should supply citations in two places in your writing. Provide an abbreviated citation, called an **in-text citation**, within your text at any spot where you're conveying information you took from an outside source. In-text citations act as tags to show your reader the exact spot in your writing where you incorporated information from an external source. You've no doubt seen in-text citations in Wikipedia articles. They show up as numbers within the text and they correspond to full citations in the **References** section at the bottom of the article.

But in-text citations are not always numbered like they are in Wikipedia articles. Sometimes they take the form of the author's last name plus the date of publication. Let's look at an example from this journal article:

> Allen, C. D., et al. (2002). Ecological restoration of southwestern ponderosa pine ecosystems: a broad perspective. *Ecological Applications*, *12*(5): 1418–1433.

Below you see an excerpt from that article, showing how the in-text citations are not numbers, but rather authors and dates:

> Alteration of stand structures and species compositions has in turn altered natural processes in Southwestern forests. Understory grasses and forbs have decreased in abundance and diversity (Covington and Moore 1994, Bogan et al. 1998), replaced by deep mats of slowly decomposing pine needles. As a result, nutrient cycling dynamics have been disrupted (White 1994) and overall biodiversity levels decreased (Allen 1998). Old-growth ponderosa pine forests have become rare (Harrington and Sackett 1992, Noss et al. 1995) and meadows have shrunk due to tree encroachment (Swetnam et al. 1999).

Let's pause for a moment and consider what authors are doing when they write and include citations to others' work. Take another look at the last sentence in the excerpt above and note there are three in-text citations:

- Harrington and Sackett 1992
- Noss et al. 1995
- Swetnam et al. 1999

The authors of this article have incorporated those in-text citations to let the reader know that *they* didn't establish that old-growth ponderosa forests are rare—rather, those facts were established in a couple of other research publications, one by the authors Harrington and Sackett, and one by Noss et al. Furthermore, the authors learned about shrinking meadows from a publication by Swetnam et al. If you read more of the excerpt, you'll see the authors are documenting lots of outside research that's established problems with ponderosa pine forests. They're doing this to set the context for their *own* research project, which will investigate ecological restoration strategies for these forests. Obviously, there'd be no need to restore these forests if there weren't problems with them, so the authors are showing there *are* problems—lots of them—but *other* researchers did the work to bring those problems to light.

Did you notice that some of the in-text citations include **et al.** and are you wondering what that means? It's an abbreviation of the Latin phrase **et alii**, which means **and others**. So, this notation indicates there were several more authors in addition to Noss, and in addition to Swetnam, but not all of them are listed because that would make the in-text citation too long and cluttered.

If you want to see the full citations for any of these in-text citations, you'd just look them up at the end of the article. As an example, here's the full citation that corresponds to the last in-text citation in that excerpt:

> Swetnam, T. W., Allen C. D. & Betancourt, J. L. (1999). Applied historical ecology: using the past to manage for the future. *Ecological Applications 9*: 1189–1206.

So now you know that full citations correspond to in-text citations but provide more detail. When you find full citations at the end of a piece of writing they'll be listed in a section that might be called any of the following or something similar:

- **References Cited**
- **Cited References**
- **Literature Cited**
- **Cited Literature**
- **Works Cited**
- **Cited Works**
- **Bibliography**
- **End Notes**

Full citations will be in numbered order if the in-text citations are numbers—again, look at any Wikipedia article to see an example. But, they'll be in alphabetical order when the in-text citations use the author/date format. The way in-text citations should look—that is, whether they're numbers, author/date notations, or something else—and the way your full references should look will depend on the **citation style guide** you are using. You may already be familiar with one or more citation style guides such as APA style, IEEE style, CSE style, etc. But don't worry if you're not familiar with citation style guides because you'll learn all about them in Chapter 31.

Just so you know, there are a few more variations in the way in-text citations can look. For example, sometimes you'll see in-text citations that have the phrase "in press" like you see at the end of this excerpt:

> The first well, GRT-1, reached a final measured depth (MD) of 2580 m at the end of December 2012 (Fig. 3). Its open-hole section crosses the Buntsandstein sandstone and the fractured Paleozoic granite (Duringer et al. in press).[1]

That means the authors are referring to a manuscript by Duringer et al. that's not published yet, but is in the process of getting published.

Also be aware you can cite content from an unpublished source, and when you do, you only need to provide an in-text citation; there's no need to provide a full citation in your references cited section. For example, if you want to incorporate into your writing information you got from a conversation with someone, you can note it in an in-text citation using the phrase **personal communication**, like you see at the end of this excerpt:

> The only breeding population of Florida panthers occurs in south Florida and during the 1980s the panther population was thought to be as low as 20–30 adults, but due to protection under the Endangered Species Act and recent recovery efforts the population is now believed to range from 100–180 panthers of breeding age [9,10]. This population growth has led to an increase in panther use of private ranchlands, resulting in an increase of verified calf (B. taurus) depredations by panthers in south Florida [10], which has raised concerns and requestions for compensation (L. Priddy, Rancher and Commissioner, Florida Fish and Wildlife Conservation Commission (FWC), personal communication).[2]

Or, if you take content from unpublished correspondence, you can note it in an in-text citation using the abbreviation **in litt.**, which is Latin for **in litteris** and translates to **in correspondence**. Here's an example:

> Species-diversity indices were compared between four habitats. This study, together with Hart (1985) and J.A. Hart (in litt.), provides a benchmark of the ecology of small mammals in the Ituri Forest during minimal human disturbance.[3]

Again, for these types of in-text citations and only these, you needn't provide a corresponding full citation at the end of your text.

Be aware that in some humanities and social science disciplines, you may find full citations in **footnotes** rather than at the end of the work in a **references** section. Footnotes appear at the bottom of a page of text where the corresponding in-text citation appears.

---

[1]Maurer, V. et al. (2020). Seismicity induced during the development of the Rittershoffen geothermal field, France. *Geothermal Energy*, 8: 1–31.

[2]From: Jacobs, C. E., & Main, M. B. (2015). A conservation-based approach to compensation for livestock depredation: the Florida panther case study. *PLoS One*, 10(9): e0139203.

[3]From: Gubista, K. R. (1999). Small mammals of the Ituri Forest, Zaire: diversity and abundance in ecologically distinct habitats. *Journal of Mammalogy*, 80(1): 252–262.

Why Should You Cite?

One reason authors consult and cite outside sources is to show readers how their research is grounded in our current knowledge of the world. In other words, citing shows how current research builds upon the findings of previous studies, and this provides readers with *context*. You saw an example of this in the previous section where authors cited research establishing problems with ponderosa pine forests so they could set the stage for their discussion about restoration strategies for those forests. Basically, those authors were showing the reader that other studies already established certain facts, and the research they intend to put forward will build upon and extend those facts.

Not only does citing provide context, it also helps authors build a case by providing evidence and substantiation. This gives writing more **credibility.** Recall from Chapter 17 that when an author cites outside sources in their writing, it's as if the author is saying: "I'm so confident in the accuracy of the information I'm writing, that I invite you to double-check my sources if you don't believe me." So citing is an act of transparency that allows readers to locate and inspect the sources an author used, in case the reader wants to verify that the author used those sources appropriately. Authors may also cite to draw readers' attention to outside sources that agree with or support their findings. This further boosts the authors' credibility because they are demonstrating to the reader that a consensus is emerging.

Sometimes authors cite publications that contain flawed findings to show how *their* research *corrects* those flawed findings. Or they may cite publications that contain findings that are *not* in agreement with their findings. One reason for doing this is to draw the reader's attention to hypotheses that may need more scrutiny, testing, and modification.

Citing is a Moral Obligation—Otherwise You're Committing Plagiarism

When you write or when you deliver a presentation, if you incorporate information you gathered from outside sources, then you are morally obliged to cite those sources. Using information from another source without acknowledging and honoring the efforts of the creator is not only unethical, it's also a form of stealing called plagiarism. **Plagiarism** is the act of taking text, words, ideas, or other content from a source without indicating your use of that source. This is also known as **not giving credit** to the source. When you present information that comes from external sources but you don't credit those sources, then you're passing that information off as your own, which is what makes this an act of fraud. Table 29.1 lists different types of content that should not be plagiarized and where that content might be found.

If you plagiarize, then you're acting in an immoral and unethical fashion. Furthermore, in some instances plagiarism is also an act of copyright infringement that may be subject to legal prosecution.

Table 29.1

Content that should not *be plagiarized*	*Sources where that content may be found*
• text • ideas • images • art • graphics/illustrations • videos • presentations	• books • newspaper articles • websites • journal articles • magazine articles • songs • artworks • television shows • YouTube videos • email messages • interviews • conversations • letters • advertisements • podcasts • computer programs

Of course, you don't need to cite any information or ideas that originated from *you*. When you write, readers will assume they are reading your ideas and your ideas only. So, there's no need to cite yourself. It's only when you refer to information from external sources that you need to alert your reader that they are encountering information that *didn't* originate with you. Nor do you need to cite what's called "common knowledge," such as that salt consists of sodium and chloride. Plus, commonly used terminology like "climate change," "post-traumatic stress disorder" or "plate tectonics" don't need to be cited because these phrases have become part of everyday language.

Let's Get Even More Specific about Plagiarism

What does it mean to take from a source in a way that constitutes plagiarism?

- Using your own words to paraphrase or summarize from a source without giving credit to the source.
- Copying whole sentences from a source without giving credit to the source.
- Copying parts of sentences from a source without giving credit to the source.
- Copying phrases from a source without giving credit to the source.
- Copying diagrams, illustrations, charts, pictures, or other visual materials without giving credit to the source.
- Copying media, including images, audio, video, or other media without giving credit to the source.

By the way, "copying" means making a *very similar* or *identical* version of something. In your writing, it's never appropriate to copy text (or other content) from a source and pass that text off as your own. Also, some students think anything they find on the internet is free to use so it's fine to copy it, and there's no need to give credit—but that's incorrect.

Copying from the internet is considered plagiarism.

What Are the Consequences of Plagiarizing?

Plagiarism is a serious transgression. Depending on the circumstances, plagiarizing could result in an "F" grade on an assignment, or it could result in flunking an entire course. Depending on how serious the offense is and/or whether plagiarism is committed multiple times, it could result in being expelled from college or graduate school. But plagiarism is just as serious after you graduate. People lose their jobs for plagiarizing, as you can see from this *The Wall Street Journal* headline:

> Mullin, B. (2020). BuzzFeed fires reporter after investigation into plagiarism. *Wall Street Journal.* https://www.wsj.com/articles/buzzfeed-fires-reporter-after-investigation-into-plagiarism-11593221507

So, be very mindful you don't plagiarize while you're in college or after!

To sum up, what should science and engineering students know about citing?

1. Citing conveys to your reader that you used an outside source, and it provides your reader with relevant details for finding and consulting that source for themselves, should they want to.
2. In your writing, you'll need to cite in two spots: within your text, using an in-text citation, and at the end of your text, using a full citation.
3. Citing enhances your credibility and transparency as a writer, but it's also a moral obligation to give credit to others.
4. When you use an outside source but don't credit it, you're committing plagiarism, which can have serious consequences.

What's Next?

In the next chapter you'll learn how to **incorporate citations** into your writing.

CHAPTER 30
INCORPORATING CITATIONS INTO WRITING

This chapter will explain different methods for incorporating outside sources into your writing. Each method serves a particular purpose, so you shouldn't use them randomly. Rather, you'll need to select the method best suited for your needs, and this chapter will provide guidance for doing that.

Directly Quoting

Let's jump right in and start with **quoting**. To **directly quote** means to take words, phrases, parts of sentences, or full sentences from another source and integrate them into your writing. When doing so, you'd use quotation marks to show what you took from the outside source. If you take multiple consecutive sentences or paragraphs, then you'd use a **block quote** instead of quotation marks.

Here is an example of a block quote from an article that appeared in *The New York Times*. In this article, a journalist interviews Jason Wright, an astronomer at Penn State University, about searching for intelligent life in the universe:

> Biosignatures, Wright said, are going to be "extremely challenging to detect—if they exist … It's very possible that life is so rare that there's nothing within a kiloparsec for us to find." But technology, he explained, could have started some distance away—a kiloparsec is 3,261 light-years in distance—and moved closer to Earth over eons. It could be a traveling probe like one of our Voyagers or a systematic species migration; it could be an electronic signal, sent 3,250 years ago and, moving at the speed of light, just coming into our range.
>
> "So we have a much bigger radius for technology," Wright said. "But also, perhaps complex life that builds technology is itself extremely rare, even when life forms." He paused. "I don't know," he said. "What drives me is not the idea that we will find something in my lifetime. What drives me is that we're not looking very well. And it's too important a search, answering too important a question, not to do well."[1]

Notice how in popular science and engineering writing, such as this newspaper article, direct quotations bring the text alive for the reader, plus quotes make the content more personal and more compelling.

---

[1]Gertner, J. (2022). The search for intelligent life is about to get a lot more interesting. *The New York Times*. https://www.nytimes.com/2022/09/15/magazine/extraterrestrials-technosignatures.html/.

Now let's look at the use of a direct quote in an essay from the journal *Conservation Biology*. In this essay the authors are arguing against using a "triage" method for conserving species (in which you'd cut your losses on those species that are hardest to save and just let them go extinct, and instead focus efforts on those that are easiest to save). Here's what they say at the end of their essay, and notice how they add power to their argument by incorporating a relevant, expressive, and persuasive quote from the famous naturalist, Aldo Leopold:

> We should be asking how we can best address the conservation needs of species, not whether the needs can be addressed. We call on conservationists to work together to understand, develop, and invest in the resources needed to maintain and recover all species, not just those for which the task may seem less challenging. Rather than settling for a flawed system of triage, striving to recover all species and prevent as many as possible from becoming threatened keeps all of the pieces of biodiversity together and follows Aldo Leopold's dictum, "To keep every cog and wheel is the first precaution of intelligent tinkering" (Leopold 1966). Extinction is irreversible, and should not—and need not—be acceptable. The explicit goal of zero extinction is a key paradigm in species conservation, one which should not and cannot be dismissed.[2]

As these two examples illustrate, when someone has said something that's particularly vivid, dramatic, or eloquent—and your reader would benefit from hearing it without any modification—that's when you use direct quotes.

You might also see direct quotes used when someone has coined a new term, or used a term or phrase in a unique or particularly expressive way, like in this excerpt:

> The degradation of fresh waters as a clean resource is discussed by Wetzel (1992) and Stiassny (in press). Roberts (1993:107) noted that whereas different chemical pollutants might, individually, be relatively innocuous to fishes, the pollutants can be lethal when found together. Roberts engagingly referred to this process as "the double whammy." Myers (1996:1029) described the same process as "synergism," that is, the uniting of energies of two discrete mechanisms such that together they have an amplified effect.[3]

Be aware that in science and engineering disciplines, you'll rarely see direct quotes used in **research articles** or **conference papers**. That's because the objective of these sources is to convey the methods and results of a research project in a neutral and concise way—so, using direct quotes to embellish the text is rarely appropriate. Instead, these types of sources rely heavily on **referring** to incorporate outside sources. You'll

[2]Wiedenfeld, D. A., et al. (2021). Conservation resource allocation, small population resiliency, and the fallacy of conservation triage. *Conservation Biology*, 35(5): 1388–1395.

[3]Harrison, I. J., & Stiassny, M. L. (1999). The quiet crisis: a preliminary listing of the freshwater fishes of the world that are extinct or "missing in action." In *Extinctions in Near Time* (pp. 271–331). Springer.

learn more about this technique shortly, but for now just note that different disciplines have their own norms and protocols for how to incorporate and cite information from external sources.

Paraphrasing

Another way to incorporate outside sources into your writing is to paraphrase information from a source. **Paraphrasing** means using your own words to express content you've borrowed from an outside source. You should paraphrase in situations where using your own words will convey that content more clearly and smoothly. Let's revisit a couple of excerpts from above that have examples of paraphrasing. In that quote about the effects of pollutants on fishes, you can see a mix of paraphrasing and direct quoting. Starting with the second sentence, content is paraphrased from the original source, but with phrases like "the double whammy" and "synergism" inserted as direct quotes. Also, if you take another look at the excerpt from *The New York Times* article, you'll see that starting with the second sentence content is paraphrased from a conversation the journalist had with the astronomer, Jason Wright.

Summarizing

A third way to incorporate outside sources into your writing is to summarize. **Summarizing** provides your reader with a brief overview of information you've taken from a source. You'd use this method when you want to provide your reader with a quick synopsis but not all the nitty-gritty details. Here's an example of providing a summary:

> R.D. Martin (1993) recently published an estimate of the time of origin of primates, based on a coalescence model with diversity increasing backward in time at a constant rate from the present to the origin of the order. This involved the average duration of species, the number of primate species living today, and the number of primate species known as fossils. Martin assumed a fixed species duration of 1 m.y., a modern diversity of 200 primate species, and a primate fossil record of 186 species ranging over the past 55 million years. From this, he estimated the coalescence of his model and time of origin of primates to be around 80 million years before present [Ma].[4]

The authors of this excerpt have provided readers with a summary of R. D. Martin's work to set the stage for their own work, which is to show how mathematical probability calculations could be used to evaluate Martin's assumptions and conclusions.

---

[4]Ginderich, P. D. & Uhen, M. D. (1994). Time of origin of primates. *Journal of Human Evolution*, 27(5): 443–445.

Let's look at another example of summarizing—this time from a book. In this excerpt, the author summarizes findings from multiple publications to make the point that more education leads to a more enlightened populace.

> Studies of the effects of education confirm that educated people really are more enlightened. They are less racists, sexist, xenophobic, homophobic, and authoritarian.[10] They place a higher value on imagination, independence, and free speech.[11] They are more likely to vote, volunteer, express political views, and belong to civic organizations such as unions, political parties, and religious and community organizations.[12] They are also likelier to trust their fellow citizens—a prime ingredient in the precious elixir called social capital which gives people the confidence to contract, invest, and obey the law without fearing that they are chumps who will be shafted by everyone else.[5]

Before moving on to **referring**, which is the fourth way of incorporating outside sources into your writing, let's pause and reflect on a few things. First, notice that quoting, paraphrasing, and summarizing can be used in combination. For example, consider the sentence below, which is pulled from that excerpt about pollutants and fish. Notice it both paraphrases and summarizes something Myers said, *and* directly quotes his use of the term "synergism":

> Myers (1996:1029) described the same process as "synergism", that is, the uniting of energies of two discrete mechanisms such that together they have an amplified effect.

Second, you might find that paraphrasing and summarizing are harder skills to master compared to slotting in a direct quote in your writing. Nevertheless, it's useful to get practice reading and comprehending content, then summarizing and/or paraphrasing it in your own words. But, if you're finding it difficult, try these tips:

[5]Pinker, S. (2019). *Enlightenment Now: The Case for Reason, Science, Humanism, and Progress.* Penguin Books.

- Read the source, then set it aside and don't look at it while you write down the most important things you remember from it.
- Then, look over what you wrote, polish it up, and look at the source again to make sure you didn't borrow phrases and/or parts of sentences (or whole sentences).
- You'll then need to insert one or more in-text citations to give credit to the source you used.

Referring

The last way to incorporate outside sources into your writing is to simply **refer** to a source to show *where* you got facts or information. You'd use this method when your reader doesn't really need to know *any* details; rather, you're simply giving them the option to find the same information that you found in that source, by providing a citation the reader can follow up on. Here's an example:

> Green vehicle routing is presented as an alternative to regular vehicle routing models. Although the minimization of objectives such as cost or distance can lead to routes with low carbon emissions, it is argued that carbon emissions also depend on the speed of the vehicle (Demir et al., 2013), the load on the vehicle (Bektas, Laporte, 2011, Kara, Kara, Yetis, 2007), the inclination (slope) of the road (Tavares, Zsigraiova, Semiao, & da Graca Carvalho, 2008), and the level of congestion (Maden, Eglese, & Black, 2010) so that the shortest or cheapest route may not be optimal with respect to carbon emissions or fuel consumption.[6]

In this example, notice how the author shows where he gathered certain facts and information about variables that contribute to vehicle carbon emissions, but he doesn't elaborate any further, he just refers the reader to the relevant sources.

Below you see another excerpt from the methods section of a journal article. Again, notice that these authors don't provide the reader with details about how they determined soil texture, they simply refer the reader to the sources they used for their methods:

> To describe variation in the soil characteristics across three mesas, soil samples were collected to a 15 cm depth, every 5 m along both 50m transects. Soil collections, along both transects, were pooled for each of the 32 sampling plots. Soil samples were dried and sieved to less than 2 mm prior to chemical and texture analysis. Soil texture was determined using the hydrometer method (Gee & Bauder 1979) and soil pH with the 1:1 soil:water method (Kalra 1995). Concentrations

[6]Turkensteen, M. (2017). The accuracy of carbon emission and fuel consumption computations in green vehicle routing. *European Journal of Operational Research*, 262(2): 647–659.

> of organic, inorganic, and total carbon were determined from a Skalar Primacs Carbon analyser (Skalar Analytical B.V., Breda, Netherlands). Total nitrogen concentrations were determined from a Skalar Primacs Nitrogen analyser (Skalar Analytical B.V., Breda, Netherlands).[7]

This way of incorporating sources is commonly used by scientists and engineers when they write up their work in research articles and conference papers. Referring to outside sources provides readers with context, and it shows how research is grounded in the findings of previous studies. In other words, referring to outside sources allows authors to show how their work builds on previously established facts, and this enhances the credibility of their research. It also offers readers full transparency about where information and facts were derived. Last, it provides a way for the authors to be succinct and concise; instead of explaining all the information they found in those outside sources, they just refer the reader to those sources.

To sum up, what should science and engineering students know about incorporating citations into their writing?

1. Use direct quotes when you want to convey information that's particularly vivid or eloquent. But be aware that it's not standard to use direct quotes when writing up the methods and results of a research project.
2. Use paraphrasing when you need to convey content from an outside source but you can do so more clearly and smoothly using your own words.
3. Summarize from sources when you need to provide your reader with a quick synopsis but not all the nitty-gritty details.
4. To provide context and credibility, yet still be concise, scientists and engineers often use referring as a method for incorporating outside sources. This method provides readers with information about where to find additional details, should they want to follow up.
5. These methods for incorporating outside sources can be used in combination.

What's Next?

In the next chapter you'll learn how to **format citations** using **citation style guides** and how to find the elements you need to populate your citations.

[7]Ware, I. M., et al. (2014). Conflicting management objectives on the Colorado Plateau: understanding the effects of bison and cattle grazing on plant community composition. *Journal for Nature Conservation*, 22(4): 293–301.

CHAPTER 31
FORMATTING CITATIONS

In this chapter you'll learn how to create citations and format them properly. To do that, you'll need to refer to a citation style guide, and you'll need to know where to find the bibliographic elements of a source (like author, title, year published, etc.) to put together a citation. This chapter will explain all that, as well as what DOIs are and where to find them. Plus, you'll learn how to leverage reference managers to autoformat citations into the style of your choice.

Format Citations Using a Citation Style Guide

A **citation style guide** is a set of formatting rules for how your in-text citations and full citations should look. Specifically, a style guide dictates whether in-text citations should be indicated with numbers or by using an author/date notation. These guides also specify the order in which elements should appear in full citations, as well as what should be capitalized, italicized, or bolded.

When you're writing a research paper for a class assignment, your instructor will usually tell you which style guide they want you to use to format your citations. Style guides commonly used in science and engineering disciplines include APA style, IEEE style, and CSE style. Again, each of these styles has its own rules for how in-text and full citations should look. For example, below you see the same journal article citation formatted in three different styles:

- **APA style**
  Zamora, A., Ferragut, V., Jaramillo, P. D., Guamis, B., & Trujillo, A. J. (2007). Effects of ultra-high pressure homogenization on the cheese-making properties of milk. *Journal of Dairy Science, 90*(1), 13–23.

- **IEEE style**

  A. Zamora, V. Ferragut, P. D. Jaramillo, B. Guamis, and A. J. Trujillo, "Effects of ultra-high pressure homogenization on the cheese-making properties of milk," *Journal of Dairy Science*, vol. 90, no.1, pp. 13–23, 2007.

- **CSE style**

  Zamora A, Ferragut V, Jaramillo PD, Guamis B, Trujillo AJ. 2007. Effects of ultra-high pressure homogenization on the cheese-making properties of milk. J Dairy Sci. 90(1):13–23.

At first glance these citations all look the same, but note that the authors are listed differently in all three. In APA style and CSE style the last name appears first, while in IEEE style the authors' initials precede their last name. Also, the year published appears at the end of the citation in IEEE style, whereas it's shown just after the authors and enclosed in parentheses in APA style. You'll see more subtle differences if you keep looking! These are the sorts of details you need to pay attention to when following a particular style guide to create citations.

Many citation styles originate from professional associations. For example, APA style was developed by the American Psychological Association, IEEE style was developed by the Institute for Electrical and Electronics Engineers, and CSE style was developed by the Council of Science Editors. Other examples of styles produced by professional associations include:

- American Sociological Association (ASA) style
- Ecological Society of America (ESA) style
- Geological Society of America (GSA) style

For some of the more commonly used styles, like APA style, you can find formatting rules in officially published handbooks or manuals. For example, all the formatting rules for APA style are found in this manual:

> American Psychological Association. (2020). *Publication Manual of the American Psychological Association 2020: The Official Guide to APA Style* (7th ed.). American Psychological Association.

You can also find citation formatting instructions on websites such as Purdue University's Online Writing Lab (better known as Purdue OWL). Or you can run Google searches to find specific formatting instructions. For example, if you need to find out how to cite a website in IEEE style, just run a Google search on **cite website IEEE** and you'll retrieve results that have instructions.

Professional associations are not the only entities that produce citation style guides, journal publishers do as well. So, you may encounter assignments where your instructor wants you to follow the same citation style used by a particular journal, such as the journal *Nature*. If that happens, then you can Google the name of the journal along with the words **citation style** to see if you can find relevant websites with tips on how to format your citations properly. Or you can examine articles that have been published in the journal to see how citations are formatted, and then just copy that same style.

For example, below you see citations 6, 7, and 8 from the references section of this article:

> Bao, J., & Bawendi, M. G. (2015). A colloidal quantum dot spectrometer. *Nature, 523*(7558): 67–70.

You can deduce that this journal uses numbers for in-text citations because the full citations are numbered—but you could also just glance at the text of the article to see how the in-text citations look. Note that authors should be listed with last name first and then their initials, the journal title is abbreviated and in italics, and the volume number is in bold, etc. So, you'd copy this same order and formatting for *your* citations if your assignment stipulated you use the same style as the journal *Nature:*

6. Gan, X., Pervez, N., Kymissis, I., Hatami, F. & Englund, D. A high-resolution spectrometer based on a compact planar two dimensional photonic crystal cavity array. *Appl. Phys. Lett.* **100**, 231104 (2021).
7. Laux, E., Genet, C., Skauli, T. & Ebbesen, T.W. Plasmonic photon sorters for spectral and plarimetric imaging. *Nature Photon.* **2**, 161–164 (2008).
8. Xu, T., Wu, Y.-K., Luo, X. & Guo, L. J. Plasmonic nanoresonators for high-resolution colour filtering and spectral imaging. *Nature Commun.* **1**, 59 (2010).

Be aware that some citation styles, and especially those developed by journals, are rather scant on examples. That means it may be difficult to find examples for how to cite certain types of sources like websites, or what to do if the website has no author or date, etc. If your instructor has specified that you use a style that doesn't have much formal documentation, and you can't figure out how to cite a source because there's no example to follow, then ask your instructor to provide guidance.

Sometimes instructors will tell you to use whatever citation style you like. If that happens, then save yourself a lot of hassle and pick a mainstream style like APA style because you'll easily be able to find formatting instructions on the internet for this style.

How to Find Citation Elements for Traditional Sources

When you need to cite a traditional source like a book, journal article, or conference paper, where do you find all the elements to put together a citation? There are several places you can look, and it's usually a good idea to check in multiple spots in case the first place you look has incorrect or missing information.

1 Check the Source Itself

Checking the source itself is usually the most dependable and accurate method for gathering citation details, although you may not always find all the elements you need. If you have a print or PDF version of a journal article or conference paper, then look for citation details on the first page. See Figure 6.6 for a refresher on how to do that. For print or electronic books, check the title page and copyright page to find the details you need to create a citation. For dissertations and theses you should find the citation details you need on the first few pages. If you don't find all the citation details you need on the source itself, then try the techniques described below.

2 Check Database Records

Database records usually contain all the information you need to cite a source. So, you can look up a source by title in a database to see if you can find it, and if you do, simply gather citation details from the database record. For example, Figure 31.1 shows a database record for a book; notice it has all the information you need to cite it.

Book

The people of Sunghir: burials, bodies, and behavior in the earlier upper paleolithic

Authors: Erik Trinkaus, A. P. Buzhilova, Mariia Borisovna Mednikova, Mariia Vsevolodovna Dobrovolskaia

Oxford; New York: Oxford University Press, 2014

Available in Book Stacks at GN772.22.R9 T75 2014

Figure 31.1

You can also look up sources in Google Scholar and if you find them, click the **Cite** link to see if Google Scholar provides the citation details you need. Suppose you wanted to try that for the book shown above. First, look up the book and note there are multiple results for it (see Figure 31.2). The first result is for the book itself, while the other results are for *reviews* of the book. As you want a citation for the book itself, you'd click the **Cite** link in that first result, which will bring up citations in several different styles.

However, in this instance, the citation information that's provided isn't correct and that's not uncommon. Often the citations Google Scholar provides contain formatting errors, but in this particular case Google Scholar has identified the publisher's name as "Human Evolution," when actually the publisher is Oxford University Press. That's why it's good to double-check citation information in multiple places!

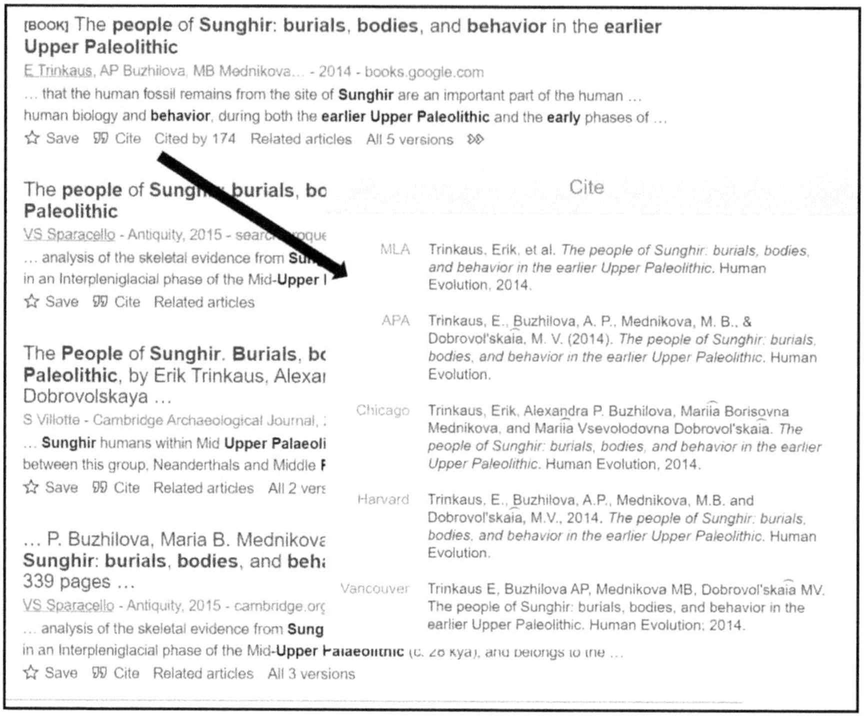

Figure 31.2

3 Google the Source

Conduct a Google search on the title of the source to see if you can find citation details for it. For example, when you do a Google search for the book, *The People of Sunghir: Burials, Bodies, and Behavior in the Earlier Upper Paleolithic,* you'll find a result leading directly to the publisher's website where you can see all the correct citation details.

Adding DOIs to Your Citations

Depending on the citation style guide you're using, it may dictate that you include DOIs in your citations, whenever DOIs are available. **DOI** stands for **Digital Object Identifier** and you can think of it as a **persistent URL** or **permalink** that'll always lead you to a specific source. Older DOIs begin with a number 10 and look like these examples:

- 10.3352/jeehp.2013.10.3
- 10.1080/02626667.2018.1560449

Newer DOIs look more like this:

- https://dx.doi.org/10.1080/02626667.2018.1560449
- https://doi.org/10.1177/0269881118806297
- https://doi.org/10.1371/journal.pone.0093354

If a source has a DOI assigned to it, you can usually find it on the first page. For example, take another look at Figure 6.6, you'll see there's a DOI just under the journal title in the header. And in Figure 7.1, you'll see the DOI in the footer of the first page of an article.

If you can't find the DOI for a source, then try looking up the source by title using Crossref (https://www.crossref.org/) to see if you can find it. If you still can't find it, then just give up because some sources simply don't have them, particularly sources published before the year 2000.

What if You Can't Find the Volume, Issue, or Page Numbers for an Article?

You may come across journal articles that don't seem to have proper page numbers, or every article in the journal is numbered from page one, or there's no volume or issue number assigned, or something else looks weird. There are several reasons why this can happen. One reason has to do with how journal publishers release new articles. Recall that most journals are published in two formats, there's an online version as well as a corresponding print version. To compile the print version, the publisher must edit all the articles that will go into an issue, assemble them together, give the issue page numbers, create a table of contents showing which page each article begins on, etc. This takes time. As this is happening, some articles that are designated for that issue will be finished and ready to go before others. For the articles that are ready to be published and just waiting for the print issue to be compiled, there's no reason not to go ahead and publish them online immediately. However, that means they may not yet have proper page numbers assigned, or even an issue or volume number, because the publisher is still working out those details. Articles published online before they come out in a print issue are variously called:

- **e-pub ahead-of-print articles**
- **ahead-of-print articles**
- **early access articles**
- **advance online publications**
- **advanced online publications**

Another reason you might encounter articles with odd citation elements is that they're published in journals that are *only* published online and have no print counterpart. For example, it's typical for online-only journals to number every article from page one. After all, when the publisher isn't bundling a bunch of articles into a physical journal issue, then there's no need to consecutively number all the pages in that issue. So, you can just begin every article on page one. A good example of this is the journal *PLoS One*—it's only published online, with no print counterpart, and every article begins on page one, as you see in Figure 7.1. Also, *PLoS One* assigns its articles a number, called an **article number**, which is used in place of the page range when you create a citation, like this:

> Mole, C., et al. (2020). Predicting takeover response to silent automated vehicle failures. *PLoS One 15*(11): e0242825.

Some citation style guides address these types of issues and provide formatting guidelines, but unfortunately many don't. So, you can try searching Google to find instructions—just enter the style you're using along with what you want to find out, like you see with these search examples:

- **APA ahead of print**
- **APA advance online publication**
- **APA article number instead of page span**

However, you may not find relevant instructions. If you don't, then you'll have to use your best judgment in compiling a citation or check with your instructor for advice.

Citation Elements for Websites

Some websites provide you with instructions for how to cite a particular page, which is always a relief. But even when that happens, you may need to reformat the citation to meet the precise guidelines dictated by your style guide. Unfortunately, for other websites you'll need to look around for the title, author, and date, and you won't always find them. Often there's no specific person designated as the author, or else there's no obvious publication date, or both! In situations like that you usually list the website owner as the author, or just start your citation with the title of the website. Consult your style guide for the appropriate protocol. If there's no publication date then many style guides dictate that you add **n.d.** to your citation, which stands for **no date**. Here's an example:

The Cornell Lab. (n.d.). *Citizen science.* https://www.birds.cornell.edu/citizenscience/

Use Reference Managers to Autoformat Citations, Including In-Text Citations

You learned about reference managers in Chapter 27 and now it's time to revisit them to let you know they can turn your stored references into a formatted bibliography. Most reference managers have lists of different citation styles to choose from to format your bibliography. Many reference managers also have a feature called a **plug-in** or **extension** that integrates with Microsoft Word to help you insert in-text citations as you write, and then generate a reference list that corresponds to those in-text citations.

Needless to say, citation autoformatting is an extremely popular feature offered by reference managers because formatting citations manually is tedious work. However, if you take advantage of this feature, be aware you still need to carefully look over the citations your reference manager autoformats to ensure there are no mistakes. There are *always* mistakes! For example, in the auto-generated bibliography in Figure 31.3, three of the four references have errors that need to be corrected, and often the mistakes are much worse than these.

While some instructors are lenient about minor mistakes in citations, others want every citation to be perfectly formatted according to the style guide they assigned. These instructors may take points off your project if your citations don't have elements in the correct order, or if you've made mistakes in punctuation, spacing, capitalization, italicizing, etc. So, you can't rely on reference managers to completely do the work of formatting for you. You'll still have to consult a style guide to ensure all your citations are correct. After all, having correct, clear citations is a courtesy to your reader, who may want to track down the publications your citations correspond to.

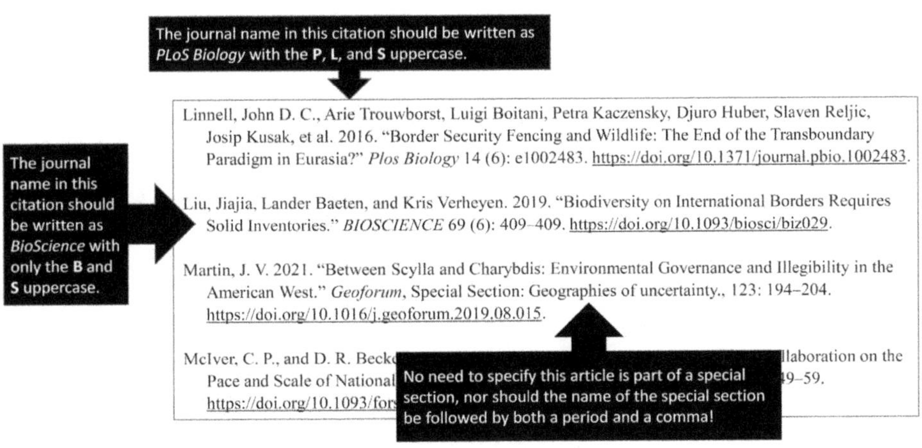

Figure 31.3

To sum up, what should science and engineering students know about formatting citations?

1. To format citations, you'll need to follow a citation style guide. These guides are provided by certain associations as well as journal publishers.
2. There are multiple places you can look for the elements that comprise a citation including the source itself, a database record for the source, or try looking the source up in Google Scholar or Google.
3. Some citation style guides require you to add DOIs to your citations when DOIs are available.
4. Reference managers can autoformat citations into a particular style, but they make mistakes, so you'll still have to double-check autoformatted citations and fix the mistakes.

What's Next?

You're done! Congratulations! If you'd like to learn about getting help, scholarly metrics, open-access publishing, or social networking, then consult the appendices that follow. But at this point you should congratulate yourself on gaining a robust set of information literacy skills!

APPENDIX A: GETTING HELP WITH YOUR RESEARCH

Most academic libraries are staffed with librarians whose job is to help students conduct research. However, typically students have no idea what librarians do, or how librarians could help them, so they don't ask for help! In fact, the two most common misconceptions students have about librarians are as follows:

- **Misconception 1**: The work that librarians do consists of shelving and organizing books, checking books out to library patrons, fixing printers, etc.
- **Misconception 2**. Librarians need expertise on the topics students are researching in order to help students find relevant sources on those topics.

What Type of Work Do Librarians Do?

Let's deal with that first misconception and set the record straight. Whereas TV and movies often depict librarians shelving books and checking books out to library patrons, in academic libraries this work is primarily done by student employees. The type of work academic librarians do includes:

- Managing the acquisition of resources, such as buying e-book packages and journal subscriptions; this work includes licensing, price negotiation, monitoring usage, etc.
- Managing library services like circulation, document delivery/interlibrary loan, course reserves, etc.
- And most importantly, helping students conduct research, including:
    - determining whether a research topic is feasible for finding relevant sources
    - assisting with the selection of keywords/synonyms for researching a topic
    - identifying the best databases to search
    - suggesting tweaks for a search that gets too many or too few results
    - helping choose the most relevant and appropriate search results, and zeroing in on sources that will meet an assignment's source requirements
    - assisting in obtaining the full text for sources
    - assisting with the use of reference management software
    - and more …

The librarians who help students conduct research often have expertise with databases and other research tools specific to a particular subject area, and because of that they're often called **subject librarians**. Subject librarians may also be referred to as **research librarians**, because they help students conduct research. Many campus libraries have subject librarians who have expertise using science and engineering databases, or social science databases, or business databases, or humanities databases, etc. Sometimes these librarians even have advanced degrees in their particular subject area. Subject librarians don't just help students with research, they help faculty too. For example, they might help a faculty member conduct a complex literature search for a systematic review article.

You may encounter subject librarians from time to time in your classroom. Sometimes instructors invite them in to teach a class session on research skills. But you can also make an appointment with a librarian to get one-on-one research help, or you can email them with questions. Most libraries provide a way to contact librarians right on the library's home page.

Librarians Don't Need Expertise on Your Topic to Help You Find Sources

Now let's address that second misconception. Librarians don't need to be experts on your research topic to help you find information sources relevant to that topic. Librarians' expertise is knowing *where* to find information sources, including which databases will be most promising to search. They're also experts at how best to *search* databases. Searching requires identifying the best keywords and synonyms to search, and this can be done *without* being an expert on the topic being researched. Librarians also know how to leverage various database search features to get the best results. Again, these skills don't require them to be experts on a topic.

So, save yourself an enormous amount of time and effort by consulting with a librarian before you start conducting research for your assignments. You'll probably find better sources and do better on your assignment!

APPENDIX B: SCHOLARLY METRICS

Anyone working toward a career in science or engineering—and especially a career that involves research—should be aware there are **metrics** that are gathered and used to judge the impact that individual journal articles as well as entire journals have on the scholarly community. Metrics are also used to measure an individual researcher's productivity and impact. While these metrics can be informative, they can also be manipulated and misused, which has made them controversial and contentious.

Scholarly metrics are also referred to as **citation metrics**, **bibliometrics**, or **scientometrics**, and there are three main types we'll examine in this appendix: article metrics, journal metrics, and author metrics.

Article Metrics

Article metrics measure the number of times an article (or conference paper) has been cited by other authors. When an article has been cited a lot—that is to say, it's a **highly cited article**—it's usually because the content and findings have been useful to lots of other researchers, otherwise those researchers wouldn't have cited it. So, in general, it's considered prestigious for an author to publish an article and then have their article cited a lot.

Keep in mind that it takes time for an article to get cited. When an article is first published, it'll have a citation count of zero until other researchers read it, use its contents, and then cite it in their publications. As time passes and more researchers use the article, the article's citation count will go up. In some cases, articles get cited quite soon after they're published, and the citation count might go up quickly. At other times, it might be years before an article gets cited by other researchers. For example, a researcher might do a survey of biodiversity in a particular region and publish their findings, but then lots of time might pass before other researchers start studying that region again. When they do, they'll likely start citing the biodiversity survey.

Be aware that if an article is highly cited, it might *not* be because it's high-quality and/or especially useful. Sometimes articles get cited a lot because they tackle controversial or contentious research topics. Or, it could be that an article is flawed and gets cited a lot as an example of a research failure (or, how *not* to do research). So, a highly cited article is not *always* an excellent article. Nevertheless, a high citation count is *usually* an indication that an article is especially useful and/or groundbreaking, so that's why it's come to be a sign of prestige.

Where to Find Article Metrics

There are several databases that provide article citation metrics. Examples include Google Scholar, Web of Science, and Scopus. To find the citation metrics for an article, pull up a record for it in one of these databases, then check the record for links with this sort of wording:

- **Citations**
- **Cited by**
- **Citing documents**

For example, in Google Scholar you'd click the "Cited by" link, which is shown in Figure 26.3 right next to the "Related Articles" link. Clicking that will bring up a list of all the articles that have cited that publication.

Journal Metrics

You learned all about journals in Chapter 6, including the fact that some are considered predatory and are *not* legitimate journals. But even legitimate journals fall along a spectrum ranging from those that are extremely prestigious to publish in to those that are less prestigious. The prestige of a journal is based on a metric that's calculated by how often articles published in that journal get cited. Journals containing articles that get cited a lot have a higher metric (and thus more prestige) than journals containing articles that are cited less. The idea here is that journals that publish lots of highly cited articles have a higher **impact** on the research community, and that's what makes them more prestigious.

The precise way journal metrics are calculated can vary a bit depending on which organization is calculating it, and each organization making the calculation has coined their own name for the metric, such as:

- **Journal Impact Factor**
- **CiteScore**
- **SCImago Journal Rank**
- **Eigenfactor**

Because time must pass before articles get cited, journal metrics are calculated by looking at articles published in a journal over the last few years (it could be anywhere from two years to five years depending on which organization is calculating the metric). This gives articles some time to get cited by other articles. Each journal's impact metric is regularly recalculated and adjusted; so, if a journal starts publishing articles that are cited less, the journal's impact metric (and thus prestige) will drop. Not surprisingly, journals that have managed to achieve a high-impact factor have editorial teams who carefully select and publish articles that look like they'll be especially groundbreaking, useful, and influential—and thus will get cited a lot. That way they can ensure the impact factor for their journal stays high and doesn't drop.

But how is it that these prestigious, high-impact journals can be so choosy and publish only the best, most interesting articles? This directly relates to the publish-or-perish requirement which you learned about in Chapters 2 and 7. In order for faculty researchers working at colleges and universities to keep their jobs, get promoted, and be competitive for other jobs, not only do they need to publish a lot, they also need to publish in the most prestigious journals possible. Doing so makes them stand out as more competitive and successful than their peers. So, if you're in academia, your reputation, job prospects, tenure and promotion prospects, and salary may depend on *how much* you publish as well as *where* you publish.

Because everyone wants to boost their career prospects by publishing their work in the most prestigious journals, those journals get thousands of article manuscript submissions. So, the editors for those journals can be very choosy and only publish manuscripts that look like they'll be cited by lots of other researchers. In other words, these prestigious journals have in effect gained a monopoly on publishing the highest-quality and most groundbreaking articles.

Journal Metrics in Relation to Journal Pricing

You just learned that prestigious journals have a monopoly on publishing the highest-quality and most groundbreaking articles. Unfortunately, monopolies offer opportunities to exploit. As prestigious, high-impact journals publish lots of top-notch, useful articles, it's very important for academic libraries to subscribe to those journals because everyone wants to read the articles published in them. That means publishers can charge very high subscription rates for their most prestigious journals, knowing that libraries will feel pressure to subscribe to those journals, no matter what the cost.

But, there's another dimension to this problem. Because academic researchers must constantly strive to publish in the most prestigious, highest-impact journals, those journals start to take on their own mystique or "academic glamour." In other words, they become *fashionable* to publish in. This means the publisher can ratchet up the subscription price even more, knowing that academic libraries will keep paying, because the journal has so much power and sway in the academic world.

Just how expensive are subscriptions to science and engineering journals? Libraries may pay more than $20,000 a year for really expensive journals, while the cheapest journals are rarely under $1,000 a year. But most libraries buy journals from publishers in package deals (called **big deals**) that include multiple journals. Journal packages typically cost anywhere from $50,000 a year and up. Academic libraries at larger institutions may pay over a million dollars a year for a journal package. For example, in 2019 University of California campuses were paying $10 million a year for a journal package from the publisher Elsevier (they canceled their contract with Elsevier that year because of the cost). When you consider that libraries are paying for *multiple* journal packages from different publishers, some need annual budgets in the multi-millions to cover the costs.

Of course, journals published by open-access publishers (which you learned about in Chapter 2) are the exception, because they are free.

Not surprisingly, when you value researchers based on how much they publish and where they publish, there are consequences. Specifically, you create a market that allows prestigious, overpriced journals (or journal packages) to flourish. The high prices of these journals have, in turn, resulted in a couple of interesting developments:

1. The open-access publishing movement started as a reaction against the exorbitantly high journal prices charged by many commercial publishers. If you recall from Chapter 2, open-access publishers charge no subscription fees for access to their journals. Instead, the author pays a fee to cover the cost of publishing their article, but the journal is free for anyone to access. You can learn more about open-access publishing in Appendix C.
2. Predatory publishers came into existence when they saw they could copy the business model of the open-access publishing movement, and prey upon authors who are anxious to publish more articles.

Note that some people mistakenly think that all open-access publishing is synonymous with predatory publishing, but it's not. Open-access publishing came about as an attempt to circumvent price-gouging from commercial publishers. Unfortunately, predatory publishers saw an opportunity to copy the business model of open-access publishers, which tainted the reputation of the open-access publishing model—at least for those who don't understand the distinctions between the two.

Additional Problems with Journal Metrics and Journal Prestige

You've seen that when researchers are compelled to publish in the most prestigious journals, that further increases the status and glamour of those journals, which motivates publishers to raise the subscription price. But that's not the only problem. In deciding where to publish a manuscript, a primary concern for authors should be "What audience will benefit most from reading my article, and what journals does that audience read?" In other words, an author's main concern should be where to publish their article so it'll be seen and used by those researchers who are most likely to benefit from it. For example, if a researcher was publishing an article on the behavior of a certain species of wasp, the best way to reach other wasp researchers might be to publish in *Journal of Hymenoptera Research* rather than publishing in the highest-impact journal possible. However, the career boost that researchers gain by publishing in the highest-impact journals usually outweighs any other considerations about where to publish.

Where to Find Journal Metrics

All the journal metric tools listed below calculate journal impact metrics in slightly different ways, but for all of them, the calculation is ultimately based on citation counts of articles published within the journal.

- **CiteScore**—This is an impact metric developed by the company Elsevier and it's available to consult if your library provides the database Scopus.
- **Eigenfactor.org**—This tool is free to use and easy to find with a Google search. You can just enter a journal title to look up its Eigenfactor score.
- **Journal Citation Reports**—This is the first and oldest tool for finding a journal's impact metric, and where the phrase "journal impact factor" was coined. Access to Journal Citation Reports requires a subscription, so check with your library to see if access is available.
- **SCImago Journal Rank**—This tool is also free to use and easy to find with a Google search.

Author Metrics

Author metrics are designed to measure a researcher's influence and productivity by running a calculation that simultaneously looks at the number of articles they've authored along with the number of times those articles have been cited. The most commonly used author metric is called the **h-index**. To learn more about it, including details about how it's calculated, just do a Google search on **h-index** and you'll find plenty of information. Another, similar statistic is the **Author Impact Beamplot**, which was developed by Clarivate, the company that produces the article database Web of Science.

Where to Find Author Metrics

There are several options for looking up an author in a database to view their metrics:

- **h-index**—This metric is available from databases like Web of Science, Scopus, and Google Scholar. However, it's only available in Google Scholar if the author has set up a Google Scholar profile. If the author has done that then you can (1) search their name in Google Scholar; (2) click on their profile to view it; (3) look at their h-index, which appears to the right of their publications on their profile page.
- **Author Impact Beamplot**—This metric is available from Web of Science.

Ensuring the Accuracy of Author Metrics with ORCID

Author metrics can be tricky to properly calculate, partly because lots of authors have the exact same name—or an author might change their name over time, perhaps because they marry and take their partner's surname. For these reasons it can be tricky to sort out exactly which articles a particular author wrote, making it difficult accurately calculate the author's h-index score.

The solution to this problem is called **author name disambiguation** and it can be achieved by having researchers register to get a unique numerical identifier called **ORCID**, which stands for **O**pen **R**esearcher and **C**ontributor **ID.** An ORCID number distinguishes a particular author from other authors, including those that share the exact same name. Registering for an ORCID also ensures an author's publications are linked to them even if they change their name and start publishing under a new name. If you want to register for an ORCID, you can easily find the ORCID site in a Google search. Once there you'll see a link to register for an ID. It's very easy to set up!

Problems with Scholarly Metrics

While scholarly metrics often factor in to hiring, promotion, and tenure decisions for researchers at academic institutions, these metrics may not be measuring a researcher's productivity, value, or impact particularly well. Not only that, it's questionable whether these metrics accurately portray how researchers measure up against each other. Consider that an author may be publishing extremely useful content, but in a very narrow research area that has few other scholars. That author's publications may never get cited a lot simply because there aren't many other scholars who will use and cite the research. That means the author will have a low h-index even though the content they're publishing is high quality.

Or, perhaps an author is publishing excellent content that has direct, real-world applications, and that content is being heavily used by practitioners and technicians, but not so much by other scholars. So, the content is highly practical and useful, but not to researchers who regularly write and cite. Authors who fall into this category will also have low h-indexes, even though their work is useful and high quality.

Another problem is that metrics can be manipulated and gamed, and you learned a bit about this back in Chapter 14. For example, authors can boost their h-index by citing their own publications. They can also manipulate their metrics through "salami publishing," which is writing up the results of a research project in as many separate articles as possible. Some journal publishers are on the lookout for these practices and will refuse to publish manuscripts from authors who engage in these behaviors—but it still happens.

Furthermore, metrics *de-incentivize* researchers from embarking on novel, highly exploratory research projects, because those projects are more likely to fail and not result in citable publications (see Chapter 7 for more information about the problems researchers have publishing negative results). On the other hand, metrics *incentivize* researchers to make the results from their research projects seem more dramatic and groundbreaking than they actually are in order to boost citations to their publications. This can be done through practices such as "p-hacking," "data-dredging," or "HARKing." You can learn about these practices with a Google search—or just look them up in Wikipedia, which provides good coverage.

It's not just authors who try to boost their metrics; journal publishers will use highly questionable tactics to boost their journals' impact metrics. For example, if a researcher gets a manuscript accepted for publication in a journal, the journal's editor might pressure the researcher to rewrite their manuscript a bit to include citations to other articles published in that journal. This "coercive citing" tactic will boost the journal's impact factor. Journals may also publish editorials that heavily cite other articles published in that journal to achieve the same effect. Or, journals might artificially delay committing their ahead-of-print articles to a print issue, as this gives the ahead-of-print article more time to be cited before it's formally published, thus increasing the journal's impact factor. Plus, as you learned in Chapters 7 and 14, most journal publishers won't publish articles with negative results because those articles are unlikely to get cited as much. Journals may also avoid publishing manuscripts describing novel, highly exploratory research projects because it may take years for other researchers to start citing that content, which is too long for those citations to figure into the journal's impact calculation.

These are just a few of the problems with scholarly metrics. While some policies and initiatives have been put in place to discourage gaming or to correct the inappropriate incentives metrics create, success has been limited. Plus, clever new ways to game the system are usually found.

Altmetrics

There's another type of metric you should know about called altmetrics. **Altmetrics** track the amount of attention a publication is getting in online media, such as Facebook, Wikipedia, Reddit, blogs, news sources, etc.

Depending on who is calculating the altmetric, this statistic may include measures of how often a publication has been viewed, the amount of discussion about a publication on social media platforms, and/or how often a publication has been saved to reference management software. The most well-known altmetric scores come from a company called "Altmetric," which provides an "Altmetric Attention Score" and a visualization "donut" to show where a publication is getting the most attention. You can visit the company's website to learn more.

Of course, altmetrics can also be gamed in a variety of ways and so they have their own set of controversies. If you're interested in altmetrics and want to learn more about them, just run a search on **altmetrics** in both Google and Google Scholar and peruse your results.

If you want to learn more about *both* scholarly metrics and altmetrics, then find a copy of this book, which provides a thorough treatment of the topic:

Vinyard, M. W., & Colvin, J. B. (2022). *Demystifying Scholarly Metrics: A Practical Guide*. Bloomsbury Publishing.

APPENDIX C: OPEN-ACCESS PUBLISHING AND THE OPEN-SCIENCE MOVEMENT

You learned a bit about open-access publishers in Chapter 2, but now let's dig a bit deeper and look at different types of open-access publishing models, including **gold**, **green**, and **hybrid**. These models pertain specifically to *journal* publishing—and there are even more models out there, such as delayed, diamond, bronze, and platinum—but we'll just discuss gold, green, and hybrid in this appendix. There are also **open-access books** and other types of open-access content, all of which is part of a broader **open-science movement**. More on that in a moment.

Gold Open-Access Publishing

In the **gold open-access** publishing model, an entire journal is open access, meaning there are no subscription fees or access fees to view articles in the journal. In this model, the author pays a fee to cover the cost of publishing their article. This fee is called a **publication fee** or **article processing charge**.

Some publishers exclusively produce open-access journals and so are called **open-access publishers**. The most prominent example is Public Library of Science (PLoS). As of this writing, article processing charges for PLoS run from $800 to $2,700 depending on which PLoS journal you're publishing in. Many authors are able to pay these fees using grant money or funds supplied by their employer.

The gold model of open-access publishing was designed to circumvent the traditional subscription-based journal publishing model and the exorbitantly high profit margins that commercial journal publishers were accumulating at the expense of academic library budgets (see Appendix B for more details). However, some open-access publishers (such as BioMed Central) have been bought out by commercial publishers, and some commercial publishers have developed a few of their own fully open-access journals.

Green Open-Access Publishing

The **green open-access** publishing model is a bit more complicated. In this model, authors negotiate with publishers to publish two copies of their article. The publisher incorporates one copy into a traditional, subscription-based journal—so this copy is not free. But the publisher allows the author to make another copy available for free. Usually the free copy is either a **pre-print**, which is the manuscript version of the article, or a **post-print**, which is the manuscript version of the article *after* it's undergone peer review, but before it's been typeset by the publisher.

The free version of the article is usually deposited by the author into a **pre-print database** or into an **institutional repository**. (See Chapter 20 for more information about pre-print databases and institutional repositories.) This activity is called **self-archiving** because the author is archiving their own copy of their article to make it freely available to others. However, sometimes the author is not allowed to self-archive the free version of their article until after a delay, which is called an **embargo**. Embargoes are often 6 months or 12 months and sometimes longer. All the details about what form the free copy of the article will take, and whether there's an embargo period, are negotiated between the author and the journal publisher.

Hybrid Open-Access Publishing

In the hybrid model of open-access publishing, authors publish just one copy of their article in a traditional, subscription-based journal, but they pay the journal publisher an additional fee (an **article processing charge**) to make that article free for anyone to read. For example, as of this writing, to publish an article in the subscription-based journal *Nature* but make the article open access will currently cost authors over $10,000 in article processing charges. So, the commercial publisher Springer, who owns the journal *Nature*, makes profits off this journal from both subscription fees as well as by collecting fees from authors who want their *Nature* article to be free for anyone to read.

By the way, it's fairly common for traditional, subscription-based journal publishers to indicate on their website that they support and offer *gold* open-access publishing, when they really mean *hybrid* open-access publishing.

The Open-Access Publishing Movement

All these new forms of open-access journal publishing came about from the late 1990s through the mid 2000s as part of an **open-access publishing movement**. This movement began as a reaction to the extraordinarily high subscription prices charged by many journal publishers, resulting in those publishers earning staggering profit margins of up to 30 percent or 40 percent. See Appendix B for more information about how high journal prices are, and why they are so high.

Sadly, despite the open-access publishing movement, journal prices remain very high to this day. Nevertheless, the movement drew attention to the problem of journal pricing and has had other positive outcomes, such as pushing forward a broader **open-science** movement, which you'll learn more about shortly.

A major milestone in the open-access publishing movement was a 2002 meeting in Budapest that launched the **Budapest Open Access Initiative**. This meeting resulted in a formal declaration promoting the idea that research results should be free and openly available to anyone.[1] The open-access publishing movement gained more momentum and traction when open-access mandates were imposed by grant-funding agencies. **Open-access mandates** require researchers to make the results of their research publicly available for free.

In the United States, the first funders to impose open-access mandates were federal grant-funding agencies such as the National Institutes of Health, the National Science Foundation, the Department of Energy, etc. These funders began imposing open-access mandates because there was a growing feeling that research funded by taxpayer dollars should be freely available to the taxpayers who funded it. In other words, if taxpayers are paying to fund research through federal grant-funding agencies, then they should not have to pay again to view the results of that research. Yet, when researchers publish their research results in journals that require a subscription, those results are not free. Open-access mandates put an end to this problem. And, as time passed, many other funders imposed open-access mandates as well.

Of course, in order for researchers to comply with open-access mandates, there must be plenty of gold open-access journals to publish in. However, there aren't many, plus researchers are not incentivized to publish in these journals because most don't have the prestige and glamour of existing subscription-based journals (see Appendix B for more details about journal prestige). Instead, authors wanted to keep publishing in established, high-status journals, and both the green and hybrid open-access publishing models allow them to do that. You may have noticed, though, that the hybrid open-access publishing model allows publishers to charge subscription fees *as well as* article processing charges, thus making even more money!

If you want to learn more about the open-access movement, then try a Google search on **open-access publishing movement**. Or, if you're interested in a specific aspect of this movement or a particular publishing model, then just run a more specific Google search, such as **delayed open access.** You may also want to consult:

Suber, P. (2012). *Open Access*. The MIT Press. https://openaccesseks.mitpress.mit.edu/

Finding Open-Access Journals and Articles

To find open-access journals and open-access journal articles, there's a dedicated database called the Directory of Open Access Journals (DOAJ). You can easily find it with a Google search to learn more about it and to search it.

The Open-Science Movement

The open-access movement stimulated a broader **open-science movement**. Two important objectives of the open-science movement are to make information freely

[1]You can read the declaration here: https://www.budapestopenaccessinitiative.org/read/.

available and to increase the transparency of the research process. To learn more, consult the **Wikipedia** article on open science (https://en.wikipedia.org/wiki/Open_science) or run a Google search on "**open science**" to find more information. Note that some of the best sources of information about open science are available here:

- **Center for Open Science**: https://www.cos.io/
- **NASA's Open-Source Science Initiative (OSSI)** https://science.nasa.gov/open-science-overview
- **NASA's Transform to Open Science (TOPS) Mission**: https://nasa.github.io/Transform-to-Open-Science-Book/index.html

Open Educational Resources

One facet of the open-science movement is the development of **open educational resources**. As a student who probably pays a lot for your course textbooks, you might be particularly interested in these resources. **Open educational resources (OER)** are teaching materials, such as textbooks, that are intentionally created to be free for anyone to access, read, and use. In fact, **OER textbooks**—also called **open textbooks**—are becoming increasing popular. Many instructors are now assigning open textbooks for their courses to help their students save money. For more information about open educational resources, including textbooks, and where to search and find these resources, see Chapter 3.

APPENDIX D: SOCIAL NETWORKING FOR SCIENTISTS AND ENGINEERS

Although many researchers use social networking platforms like Facebook and LinkedIn, there are also social networking platforms designed *especially* for researchers. These are often referred to as **academic social networking** sites or platforms, and they allow researchers to set up profiles showing their degrees earned, institutional affiliations, contact information, research interests, publications, etc. Researchers who have publications that are open access (see Appendix C) will often load PDFs of those publications onto their profile page so anyone can easily download them. Also, depending on the features offered by the social networking site, you may be able to follow particular researchers, look up topics to see who is actively conducting research on those topics, form networks of colleagues, etc.

Examples of social networking platforms for researchers and academics include ResearchGate and Academia.edu, but you can find additional options by conducting Google searches such as:

- **academic social networking**
- **(engineering OR engineers) ("social networking" OR "social networks")**
- **scientists ("social networking" OR "social networks")**

Furthermore, some *disciplines* have their own social networking sites. To find out what's available for your discipline, just run a Google search for your disciplinary area like you see in the examples below:

- **psychology ("social networking" OR "social networks")**
- **computer science ("social networking" OR "social networks")**

Last but not least, some social networking platforms cater to a specific need or function. To find them, just try Google searches on the function you are looking for. For example, software developer and engineers engage in a lot of code sharing. To find the websites where this takes place, simply do a Google search on **code sharing (sites OR websites OR platforms)** to see what you can find.

COPYRIGHT ACKNOWLEDGMENTS

Figures

Figures showing Google and Google Scholar results: Google and the Google logo are trademarks of Google LLC.

Figures showing data from Web of Science, provided by Clarivate: Web of Science and Clarivate are trademarks of their respective owners and used herein with permission.

1.1 *New York Times*: ID 98510383 © Deanpictures | Dreamstime.com; Daily News ID 159699384 © Erin Alexis Randolph | Dreamstime.

7.1 Found at https://journals.plos.org/plosone/article?id=10.1371/journal.pone.0242825. Public Library of Science (PLoS) used with CC BY license. Accessed 1/18/2024.

7.2 Based on "How to Read (and Understand) a Social Science Journal Article" by Frederique Laubepin, PhD. Available at https://online225.psych.wisc.edu/wp-content/uploads/225-Master/225-UnitPages/Unit-05/Laubepin_IUCPSR_2013.pdf. Used with permission of the author.

10.3 Shows search results from ProQuest Agricultural and Environmental Science Collection.

12.1 Displays a data set from UC Irvine Machine Learning Repository. https://archive.ics.uci.edu/dataset/165/concrete+compressive+strength. Accessed 1/18/2024.

12.2 Found at https://journals.plos.org/plosone/article?id=10.1371/journal.pone.0225713#sec021. Public Library of Science (PLoS) used with CC BY license. Accessed 1/18/2024.

12.3 Found at https://journals.plos.org/ploscompbiol/article?id=10.1371/journal.pcbi.1007811. Public Library of Science (PLoS) used with CC BY license. Accessed 1/18/2024.

13.1 Shows an inactive patent found at https://image-ppubs.uspto.gov/dirsearch-public/print/downloadPdf/8192698.

13.2 Reproduced from https://www.ecfr.gov/current/title-23/chapter-II/subchapter-B/part-1235/section-1235.3.

17.2 Shows pages from U.S. Department of the Interior. (2021). *Invasive Species Strategic Plan*. https://www.doi.gov/sites/doi.gov/files/doi-invasive-species-strategic-plan-2021-2025-508.pdf

Cartoons

1.1 Two women and a child in a room, one woman seated watches an army of men pass the window, the child stands by the window and the other woman covers her face with her hand. Engraving. Wellcome Collection. Public Domain Mark. *Source*: Wellcome Collection. https://wellcomecollection.org/works/s5ejkjht.

2.1 A seated man reading a newspaper in a kitchen (?), a basket by his feet. Engraving by C. W. Sharpe after T. S. Goode. Wellcome Collection. Public Domain Mark. *Source*: Wellcome Collection. https://wellcomecollection.org/works/xsx2wvfb

3.1 Image from page 584, Henry Wadsworth Longfellow, in *Portrait Gallery of Eminent Men and Women of Europe and America Embracing History, Statesmanship, Naval and Military Life, Philosophy, The Drama, Science, Literature and Art with Biographies*, Volume 2 by Evert A. Duyckinck. New York: Johnson, Wilson and Company, 1873. https://commons.wikimedia.org/wiki/File:Portrait_gallery_of_eminent_men_and_women_of_Europe_and_America_-_embracing_history,_statesmanship,_naval_and_military_life,_philosophy,_the_drama,_science,_literature_and_art,_with_biographies_(1873)_(14588032449).jpg#filelinks

4.1 An illustration from *Dick Sand, A Captain at Fifteen* by Jules Verne, painted by Henri Meyer, circa 1877–78. https://commons.wikimedia.org/wiki/File:%27Dick_Sand,_A_Captain_at_Fifteen%27_by_Henri_Meyer_041.jpg

5.1 "General George Crook on the Trail." Engraving of General George Crook on horse patrol in Indian country circa 1867; published in *The Century Magazine* (Vol. XLI, No. 5) in March 1891. http://digital.library.cornell.edu/cgi/t/text/pageviewer-idx?c=cent;cc=cent;rgn=full%20text;idno=cent0041-5;didno=cent0041-5;view=image;seq=655;node=cent0041-5%3A1;page=root;size=s;frm=frameset.

6.1 Lady Morgan aka Sydney Owenson, Irish novelist, 1776, April 14, 1859. Stipple and line engraving by Robert cooper, after Samuel Lover. 1825. http://brn227.brown.wmich.edu:7000/3325/

7.1 Illustration from chapter 8 of *My Lady of Orange* (1901), a historical fiction thriller of the siege of Netherlands by Alva. Text by H. C. Bailey, illustrated by G. P. Jacomb-Hood. 1901. https://archive.org/details/myladyoforange00bailiala

8.1 Engraving by John Dixon of a portrait of Mary Bertie, Duchess of Ancaster and Kesteven by Joshua Reynolds. John Dixon, Public domain, via Wikimedia Commons. https://commons.wikimedia.org/wiki/File:Mary,_Duchess_of_Ancaster_and_Kesteven.jpg

9.1 "Outdoor Sports" found in *Twentieth Century Culture and Deportment or The Lady and Gentleman at Home and Abroad* by Maud C. Cooke. National Publishing Company, Philadelphia, PA, 1809, page 311. https://commons.wikimedia.org/wiki/File:Twentieth_century_culture_and_deportment,_or,_The_

lady_and_gentleman_at_home_and_abroad_-_containing_rules_of_etiquette_for_all_occasions_(1899)_(14778829982).jpg

10.1 Seated Man with Quill in his Right Hand at Desk Littered with Books. Célestin Nanteuil (French (born Italy), Rome 1813–1873 Bourron-Marlotte), 1834. Etching. The Elisha Whittelsey Collection, The Elisha Whittelsey Fund, 1949. https://www.metmuseum.org/art/collection/search/812506

11.1 Two women reprimanding two children. Engraving by J. Chevillet, 1785, after J. A. de Peters. Wellcome Collection. Public Domain Mark. *Source*: Wellcome Collection. https://wellcomecollection.org/works/jr7pp7q3

12.1 An alchemist peacefully writing in a room strewn with papers. Engraving by V. A. L. Texier after F. Giani after T. Wyck. Wellcome Collection. Public Domain Mark. *Source*: Wellcome Collection. https://wellcomecollection.org/works/y873ctep

13.1 From "Peasant Proprietorship," in *Once a Week: An Illustrated Miscellany of Literature, Art, Science, & Popular Information*, Volume II, December to June, 1860. London: Bradbury & Evans, page 165. Frederick Walker (1840–1875), Public domain, via Wikimedia Commons.

14.1 Engraving representing Anton Felkel between books and other tools. Unknown author. Circa 1762. https://commons.wikimedia.org/wiki/File:Anton_Felkel_between_books_-_engraving.jpg

15.1 The engraving shows a long-haired woman (Áslaug a.k.a. Kráka a.k.a. Randalín) clad in a net sitting with a dog. In the bottom left corner there is the signature of Mårten Eskil Winge (1825–1896), the artist. In the bottom right corner there is the signature of Wilhelm Meyer (1844–1944), the xylographer. Mårten Eskil Winge (1825–1896) and Wilhelm Meyer (1844–1944) as described above. Photograph by User:Haukurth, image processing by User:Bloodofox. Public domain, via Wikimedia Commons.

16.1 Russian Way of Travel. From Lacroix, Frédéric (d. 1864). *Les mystères de la Russie: Tableau politique et morale de l'empire Russe*. Paris, Pagnerre, 1845. https://commons.wikimedia.org/wiki/File:Russian_way_of_travel_(Geoffroy,_1845).JPG

17.1 Portrait of an unknown writing man, Friedrich Fleischmann, 1801–1834. https://www.rijksmuseum.nl/nl/collectie/RP-P-1915-1105. Rijksmuseum, CC0, via Wikimedia Commons.

18.1 Fig. 95—Prince Rupert. Portrait of Himself, page 264 in *A Short History of Engraving & Etching for the Use of Collectors and Students with Full Bibliography Classified List and Index of Engravers* by A. M. Hind. London: Archibald Constable & Co., Ltd., 1908. https://commons.wikimedia.org/wiki/File:A_short_history_of_engraving_(and)_etching_-_for_the_use_of_collectors_and_students;_with_full_bibliography,_classified_list_and_index_of_engravers_(1908)_(14780112181).jpg

19.1 *The Honorable Henry Fane (1739–1802) with Inigo Jones and Charles Blair.* Sir Joshua Reynolds (British, Plumpton 1723–1792 London). Oil on canvas.

Gift of Junius S. Morgan, 1887. https://www.metmuseum.org/art/collection/search/437449

20.1 *Design for a Book Illustration: Two Women Dancing around a Plant*, drawing, anonymous, German, 19th century (MET, 52.567.5). The Elisha Whittelsey Collection, The Elisha Whittelsey Fund, 1952. https://www.metmuseum.org/art/collection/search/335768

21.1 Chap. XV: Ecrivons à la Vielle (Writings for the elderly), in *Un an de la vie d'un jeune homme* (*A Year in the Life of a Young Man*), 1824. Artist Victor Adam. Publisher Sazerac and Duval, Paris. The Elisha Whittelsey Collection, The Elisha Whittelsey Fund, 1962. https://www.metmuseum.org/art/collection/search/698509

22.1 *Soldier Playing the Theorbo*. Ernest Meissonier, 1865. Oil on wood. Bequest of Martha T. Fiske Collord, in memory of her first husband, Josiah M. Fiske, 1908. https://www.metmuseum.org/art/collection/search/437050

23.1 Fig. 84, "Shipwrecked. Engraved by Frank French," from *A History of Wood-Engraving by George Edward Woodberry*. New York: Harper & Bros., 1883, page 196. https://upload.wikimedia.org/wikipedia/commons/9/90/History_of_Wood_Engraving_0198.jpg

24.1 *Farmer with chicken*, Cornelis Bolemaert (II), after Hendrick Bloemaert, 1627–1630. RP-P-BI-1434. https://www.rijksmuseum.nl/nl/collectie/RP-P-BI-1434

25.1 Portrait of mathematician George Boole (1815–1864). *The Illustrated London News*, January 21, 1865. Unknown author. https://commons.wikimedia.org/wiki/File:Portrait_of_George_Boole.png

25.2 *A bedridden sick young woman being examined by a doctor, accompanied by her anxious parents*. Engraving by F. Engleheart, 1838, after Sir D. Wilkie. Wellcome Collection. Public Domain Mark. *Source*: Wellcome Collection. https://wellcomecollection.org/works/rb5wm8hm

26.1 "Virgil, Horace and Varius, at the House of Maecenas," from *Great Men and Famous Women: A Series of Pen and Pencil Sketches of the lives of more than 200 of the Most Prominent Personages in History*, Volume 7 by Charles Francis Horne. New York: Selmar Hess, 1894.

27.1 *Portrait of Joachim Patinir*, anonymous, after Albrecht Dürer, 1530–1550. RP-P-OB-4589. https://www.rijksmuseum.nl/nl/collectie/RP-P-OB-4589

28.1 "Wilson W. Brown, Engineer. From a war-time photograph," from *Daring and Suffering: A history of the Andrews Railroad Raid into Georgia in 1862* by William Pittenger. New York: The War Publishing Co., 1887, page 36.

29.1 Illustration from *The Captain of the Nine* by William Heyliger. New York: D. Appleton and Company, 1912, page 139.

30.1 François Vincent Raspail. Line engraving. R. Burgess, Portraits of doctors & scientists in the Wellcome Institute, London 1973, no. 2441.7 Wellcome Collection. Public Domain Mark. *Source*: Wellcome Collection. https://wellcomecollection.org/works/enn45q4w

31.1 Nach dem Oelgemälde von Hermann Kaulbach. Image from page 028 of journal *Die Gartenlaube*, 1887. Der Herzensvertraute. Leipzig, 1887. https://commons.wikimedia.org/wiki/File:Die_Gartenlaube_(1887)_b_028.jpg

31.2 "Ruddy Duck, 1836," Robert Havall after John James Audubon. *The Birds of America: Plate CCCXLIII*. 1945.8.343. Gift of Mrs. Walter B. James. https://www.nga.gov/collection/art-object-page.32484.html

ABOUT THE AUTHOR

Mary DeJong is a librarian at Northern Arizona University's Cline Library. She holds a BA in History from State University of New York in Geneseo and an MLS from Rutgers University. She has over twenty-five years of experience as a librarian in research institutions (The American Museum of Natural History Library and Memorial Sloan Kettering Cancer Center Library) as well as educational institutions (George Washington University, Vassar College, and Northern Arizona University).

INDEX

www.ingramcontent.com/pod-product-compliance
Ingram Content Group UK Ltd.
Pitfield, Milton Keynes, MK11 3LW, UK
UKHW021849090226
467839UK00002B/54